AF831986

The Urban Book Series

Editorial Board

Fatemeh Farnaz Arefian, University of Newcastle, Singapore, Singapore;
Silk Cities & Bartlett Development Planning Unit, UCL, London, UK
Michael Batty, Centre for Advanced Spatial Analysis, UCL, London, UK
Simin Davoudi, Planning & Landscape Department GURU, Newcastle University, Newcastle, UK
Geoffrey DeVerteuil, School of Planning and Geography, Cardiff University, Cardiff, UK
Andrew Kirby, New College, Arizona State University, Phoenix, AZ, USA
Karl Kropf, Department of Planning, Headington Campus,
Oxford Brookes University, Oxford, UK
Karen Lucas, Institute for Transport Studies, University of Leeds, Leeds, UK
Marco Maretto, DICATeA, Department of Civil and Environmental Engineering, University of Parma, Parma, Italy
Fabian Neuhaus, Faculty of Environmental Design, University of Calgary, Calgary, AB, Canada
Vitor Manuel Aráujo de Oliveira, Porto University, Porto, Portugal
Christopher Silver, College of Design, University of Florida, Gainesville, FL, USA
Giuseppe Strappa, Facoltà di Architettura, Sapienza University of Rome, Rome, Roma, Italy
Igor Vojnovic, Department of Geography, Michigan State University, East Lansing, MI, USA
Jeremy W. R. Whitehand, Earth & Environmental Sciences,
University of Birmingham, Birmingham, UK

The Urban Book Series is a resource for urban studies and geography research worldwide. It provides a unique and innovative resource for the latest developments in the field, nurturing a comprehensive and encompassing publication venue for urban studies, urban geography, planning and regional development.

The series publishes peer-reviewed volumes related to urbanization, sustainability, urban environments, sustainable urbanism, governance, globalization, urban and sustainable development, spatial and area studies, urban management, transport systems, urban infrastructure, urban dynamics, green cities and urban landscapes. It also invites research which documents urbanization processes and urban dynamics on a national, regional and local level, welcoming case studies, as well as comparative and applied research.

The series will appeal to urbanists, geographers, planners, engineers, architects, policy makers, and to all of those interested in a wide-ranging overview of contemporary urban studies and innovations in the field. It accepts monographs, edited volumes and textbooks.

Now Indexed by Scopus!

More information about this series at http://www.springer.com/series/14773

Biljana Arandelovic · Milena Vukmirovic

Belgrade

The 21st Century Metropolis of Southeast Europe

Springer

Biljana Arandelovic
Faculty of Civil Engineering
and Architecture
University of Niš
Niš, Serbia

Milena Vukmirovic
Faculty of Forestry
Department of Landscape
Architecture and Horticulture
University of Belgrade
Belgrade, Serbia

ISSN 2365-757X ISSN 2365-7588 (electronic)
The Urban Book Series
ISBN 978-3-030-35069-7 ISBN 978-3-030-35070-3 (eBook)
https://doi.org/10.1007/978-3-030-35070-3

© Springer Nature Switzerland AG 2020
This work is subject to copyright. All rights are reserved by the Publisher, whether the whole or part of the material is concerned, specifically the rights of translation, reprinting, reuse of illustrations, recitation, broadcasting, reproduction on microfilms or in any other physical way, and transmission or information storage and retrieval, electronic adaptation, computer software, or by similar or dissimilar methodology now known or hereafter developed.
The use of general descriptive names, registered names, trademarks, service marks, etc. in this publication does not imply, even in the absence of a specific statement, that such names are exempt from the relevant protective laws and regulations and therefore free for general use.
The publisher, the authors and the editors are safe to assume that the advice and information in this book are believed to be true and accurate at the date of publication. Neither the publisher nor the authors or the editors give a warranty, expressed or implied, with respect to the material contained herein or for any errors or omissions that may have been made. The publisher remains neutral with regard to jurisdictional claims in published maps and institutional affiliations.

This Springer imprint is published by the registered company Springer Nature Switzerland AG
The registered company address is: Gewerbestrasse 11, 6330 Cham, Switzerland

Summary of the Book

Through a comprehensive study of the city of Belgrade, this book offers insights into a wide range of topics, including urban, political and planning history, place branding and creative cities, public art, the development of a post-socialist city, the creation of pedestrian zones, and the current riverfront development. This book presents Belgrade, reviewing its recent and historical developments.

The book has eight chapters, each dedicated to a specific topic. Thus, the chapter (Chap. 1) entitled "The Urban, Political, and Socioeconomic Rise and Fall of Belgrade Through Its History" introduces Belgrade's long and complicated history. This chapter is followed by the next topic (Chap. 2), "Urban Plans and the Transformation of Belgrade Through History". Belgrade has always been a border city between the East and West, a fact which has shaped its urban structure. Belgrade was oriental in character before it became transformed into the modern European city it is today.

The book continues with a chapter (Chap. 3) on one of the major urban projects in the former Yugoslavia, the construction of New Belgrade, its development and results, entitled "The Rise and Fall of New Belgrade with Current Developments in Retrospect". New Belgrade is a modern city within today's metropolis of Belgrade, built in the second half of the twentieth century. The initial concept and idea of New Belgrade was as the modern seat of political power of the new federal administrative structure for the Federal Republic of Yugoslavia.

"Pedestrian-Friendly Belgrade" is a chapter (Chap. 7) that presents an overview of past and current efforts to apply the principles of sustainable development and create pedestrian-friendly areas in Belgrade in the area defined as Belgrade Central Zone. "Belgrade's Waterfront Revitalization" chapter (Chap. 4) deals with the waterfront revitalization, presenting built and unbuilt projects that are part of the city's eternal ambition to get down to its rivers, the Danube and Sava.

The book also explores another leading contemporary topic in chapter (Chap. 5) "Reimaging Belgrade: The Case of Savamala District". Belgrade's Savamala district has transformed over the last few years into an energetic creative urban district where the old buildings in this historical neighborhood are gaining a new role of reimaging Belgrade.

"Public Art in Belgrade" chapter (Chap. 6) presents the sculptural masterpieces displayed in Belgrade. It begins with the Belgrade public artworks of Ivan Meštrović, followed by the sculptural collection in front of the Museum of Contemporary Art. Three projects that have not yet been fully implemented are presented in the second part of the chapter: The Kalemegdan Bridge project, a planned memorial for Zoran Djindjić and the most controversial, the Stefan Nemanja monument.

The book ends with the chapter (Chap. 8) "Belgrade's Development Visions and Aims". Belgrade is a modern Southeast European city that is striving to become a real metropolis and a regional leader. This chapter reveals current urban developments and explains the ambitions and long-term visions for Belgrade.

Contents

About the Authors

Dr. Biljana Arandelovic is Assistant Professor in the History and Theory of Art and Architecture and a visual artist. She holds a Ph.D. in Technical Science in architecture from TU Graz (2008) and was a Postdoc Research Fellow at UdK Berlin at the Institute for History and Theory of Design (2011/2012). She was also a visiting researcher at ASK Centre for Research on Management and Economics of Arts and Culture Institutions at Bocconi University (2016). She is a researcher in the national scientific project *The optimization of architectural and urban planning and design in the sustainable development in Serbia.*

The main focus in Arandelovic's research is the theory of contemporary cities and Art in Public Space. Since 2009, she has been teaching Contemporary Architecture, Art and Architecture and Serbian Architecture in the twentieth century at Niš University. She has taught at the Faculty of Applied Arts at the University of Arts Belgrade (2017), at the Department of Architecture, Unitec Institute of Technology (2012) (Auckland, New Zealand), and at the Art and the City summer schools in Vienna and Graz (2013/2015) where was also a program director.

Arandelovic is the author of the two books: Public Art and Urban Memorials in Berlin (Springer) and Visual Impressions: Architecture and Public Art in Graz (Leykam).

Since 2005, she has worked as a visual artist, and was represented by Marion Fischer art moments Gallery (2005/2011), where she had solo exhibitions in Graz. She uses photography as an addition to her primary practice of painting. Additionally, her work embraces urban photography in parallel with her research of contemporary cities.

Dr. Milena Vukmirovic is Assistant Professor in the Planning and Design Department within the Department of Landscape Architecture at the Faculty of Forestry, University of Belgrade. She is also a Research Associate at Urban Laboratory in Belgrade. From 2015 to 2019, she was an Adviser to the Director of Urban Planning of the City of Belgrade in the domain of public space design. Her research is focused on the relationship between the quality of public spaces and the competitiveness of cities in the built and natural environment. She is founder of

CSO 5km/h, which aims to promote walking as a sustainable mode of transport, co-founder of the Urban Laboratory Belgrade professional association, and co-founder and Executive Coordinator of the International Academic Conference on Places and Technologies. She is currently engaged in the national research project *Modernization of the Western Balkans* and is an external associate in the international projects *DANUrB* and *CLEVER city.*

Chapter 1
The Urban, Political, and Socioeconomic Rise and Fall of Belgrade Through Its History

Biljana Arandelovic

Abstract At the beginning of the twentieth century, Belgrade only had about 69,100 inhabitants, while today it has around 1.6 million inhabitants, which is 23% of Serbia's population. Belgrade has been the scene of social revolutions which have triggered frequent changes in its social and economic stratification, reflected in international relationships that have often broken down in frequent political crises, wars, and migrations. Immediately upon the city gaining autonomy, Belgrade's history saw great changes in the nineteenth century. Belgrade, within a span of only several decades, significantly changed its appearance: from a derelict, poorly arranged small town with winding, narrow streets, it started to become the principal economic, cultural, and governmental center of the young Serbian state. In the aftermath of WWI, in 1918, Belgrade became the capital of the newly formed Kingdom of Serbs, Croats, and Slovenes, but changed its name in 1929 to the Kingdom of Yugoslavia. Socialist Yugoslavia was formed from the remains of the Kingdom of Yugoslavia in the aftermath of WWII. After Tito rejected making Yugoslavia dependent on the Soviet Union, Yugoslavia left the Soviet bloc in 1948 and changed its political course by establishing a new position between the two blocs. The country took part in establishing the Non-Aligned Movement, which positioned Yugoslavia in the major global political arena. The collapse of the Yugoslav federation in the 1990s was a result of complex internal and external geostrategic circumstances, culminating in violent political conflict. Belgrade arrived in the twenty-first century in a worse state than anyone could have ever possibly predicted in the time of its glory after WWII. This chapter analyzes the urban, political, and socioeconomic rise and fall of Belgrade through its long history. This extensive overall study of Belgrade is divided into five periods: Development of Belgrade to 1914, Belgrade between two World Wars, Belgrade from the end of WWII until 1991, Belgrade from 1991 to 2000, and Belgrade at the beginning of the twenty-first century.

Keywords Belgrade · Yugoslavia · Social history · European city · Political history · Architectural history · Urban development

© Springer Nature Switzerland AG 2020

B. Arandelovic and M. Vukmirovic, *Belgrade*, The Urban Book Series,
https://doi.org/10.1007/978-3-030-35070-3_1

1.1 Introduction

Belgrade is a special city in many ways, and it has a long and often complicated history. It is one of the oldest cities in Europe according to archaeological excavations, dating from the Paleolithic period. Although Belgrade hasn't changed its physical position, it has been the capital of a number of states over time. After being the capital of the Kingdom of Yugoslavia (1918–1945), Belgrade later became the capital of the Socialist Federal Republic of Yugoslavia (1945–1992), then the Federal Republic of Yugoslavia (1992–2003), Serbia and Montenegro (2003–2006), and Serbia (2006–present).

Belgrade has gone through a turbulent evolutional period in the last 200 years of its history. Over this time, it has grown from a small town in the nineteenth century to the Southeast Europe metropolis of nearly two million people that it is today. It has passed through different stages with different forms of government, as well as different economic, social, and political systems. The struggle for independence has caused the arrival of the modernization and Europeanization of Serbian society, which has been constantly progressing since the beginning of the nineteenth century. The complete liberation from the Ottoman Empire in the second half of the nineteenth century was a long process, after two uprisings, which led to the formation of the first modern Serbian state.

Throughout Serbia's history, Belgrade has always played an important international role, but mainly due to its geostrategic position in Europe. If we look at the last two centuries of history, Belgrade was an important border town between two powerful empires, the Austro-Hungarian and Ottoman Empires, and often suffered the consequences of its geographical position. In the twentieth century, Belgrade had a new role as the capital of the new state of Yugoslavia, while the century ended in conflicts and crises.

The modern history of Belgrade has been marked with clear and deep discontinuities that have more than just hampered the stability of its development. Belgrade has been the scene of social revolutions which have triggered frequent changes in its social and economic stratification, reflected in international relationships that have often broken down in frequent political crises, wars, and migrations. Belgrade arrived in the twenty-first century in a worse state than anyone could have ever possibly predicted in the time of its glory after WWII. Illegal construction spread not just on the periphery but also in the city center. Belgrade started the twenty-first century with sheer ugliness, as well as diverse problems in many locations. The discontinuity in Belgrade's development is still clearly visible through the many incomplete areas throughout the city, as a straightforward result of the frequent periods of uncertainty and wars (Arandelovic et al. 2017).

Belgrade, the capital of Serbia, is a Southeast European metropolis, with a functional area that covers 7.4% of Serbian territory. The total population of Belgrade according to the 2011 Census is 1,659,440, though a 2017 estimate shows an increase, with a total of approximately 1,803,000 inhabitants. Around a quarter of the population of Serbia lives in Belgrade (25.4%). The metropolitan area of

Belgrade is composed of 17 municipalities (Barajevo, Voždovac, Vračar, Grocka, Zvezdara, Zemun, Lazarevac, Mladenovac, Novi Beograd, Obrenovac, Palilula, Rakovica, Savski Venac, Sopot, Stari Grad, Surčin, Čukarica), according to the Charter of the City of Belgrade in 2008. Ten of these municipalities are located in urban areas, while the other seven are suburban.

The largest share of exports from the Belgrade metropolitan region goes to Central and Eastern European Countries (Institute for Informatics and Statistics 2018).

Belgrade is the third largest city on the Danube (Figs. 1.1 and 1.2) and the fourth largest city in Southeast Europe. It is uniquely positioned on the geographical border between the Balkan Peninsula and Central Europe, as well as at the confluence of the Danube and Sava. The terrain in northern and western Belgrade is flat, while in the southeast it is hilly. There are 37 protected natural areas in the urban area of Belgrade, with most of them being biodiversity locations (Petrović et al. 2017).

Belgrade is located about 80 km from Novi Sad, which is the second largest city in Serbia, while several smaller cities such as Indjija, Ruma, Nova Pazova, and Stara Pazova are located between Belgrade and Novi Sad. It has an important geostrategic position in Europe, located within River Danube Corridors 7 and 10. In

Fig. 1.1 View of the confluence of the Sava and Danube Rivers from Kalemegdan Fortress. Until 1521, the history of Belgrade's urban development as well as the history of Belgrade itself was, in fact, the history of its fortress. The first fortress on the hill above the confluence of the Sava and Danube Rivers was built by the Romans, but even then, people were already living in this location. Photographer Biljana Arandelovic. Rights reserved

Fig. 1.2 The banks of the Sava River. Photographer Biljana Arandelovic. Rights reserved

addition to this, it is now working to position itself on the two core network corridors, the Mediterranean and Orient/East Mediterranean Corridors, as defined by the EC, which were introduced to facilitate the coordinated implementation of the core network (Arandelovic et al. 2017). In this way, the two European transport corridors promote Belgrade into the position of an important transportation hub in this part of Europe (Ratkaj 2002).

As the capital city of Serbia, Belgrade is an important regional center with developed scientific and cultural institutions, a modern Southeast European city that is striving to become a real metropolis and a regional leader (Figs. 1.3, 1.4, 1.5, 1.6, 1.7, and 1.8). Significant cultural and educational institutions were founded in Belgrade in the nineteenth century, among which is the University of Belgrade, founded in 1808 as the High School, considered one of the largest and most prestigious universities in Southeast Europe, with 31 faculties. Additionally, Belgrade University of Arts is another institution with an international reputation, founded in 1957 as the Academy of Arts to unite four academies, which today are the faculties of music, fine arts, applied arts, and drama.

The National Museum of Serbia was opened in 1844 and is the largest and oldest museum in Belgrade. Located in the central zone of the old town in Republic Square, it houses a permanent exhibition that covers a chronological period from the Paleolithic era to the twentieth century. The museum was closed for 15 years (2003–2018) due to the building's deterioration and lack of funding for renovation, followed by many controversies regarding the renovation process. The National Theatre, founded in 1869 and located across the street from the museum, forms an

Fig. 1.3 Vuka Karadžića Street in the historical city center. Photographer Biljana Arandelovic. Rights reserved

architectural assembly with the National Museum building (Fig. 1.9). Third among the most important cultural institutions is the Museum of Contemporary Art, established in 1965 and located in New Belgrade, which was also closed for a decade due to reconstruction and reopened in 2017. The museum collection holds Yugoslavian and Serbian art from the twentieth century. It is located in a building that is an architectural masterpiece designed by famous architects Ivana Antić and Ivanka Raspopović.

Belgrade is the most visited tourist destination in Serbia; in fact, the Statistics Office reports that tourists spent 2,190,474 nights there in 2017 compared to 765,087 in 2001 (Institute for Informatics and Statistics 2018).

When it comes to specific tourist attractions in Belgrade, it is evident that the number of material remains of the past, such as cultural and historical monuments, monumental buildings, and squares, is not very significant. The physical destruction of the city throughout history has led to this situation, making its tourist offer of this kind a little lower than in some other metropolises (Čomić and Vićić 2013).

Five major attacks on Belgrade in the twentieth century led to large losses. The first two offensives, led by the Austro-Hungarian army, happened in 1914 and 1915, when about 50% of the existing housing stock at the time was destroyed. The third happened on April 6, 1941, when Nazi bombardiers released 360 tons of bombs over a period of 2 days, during which 714 houses were devastated, while 22,000 were damaged. This does not even include the irreplaceable cultural loss for Serbian society, namely, the burning down of the National Library (Fig. 1.10),

Fig. 1.4 Oblićev Venac in the historical city center. Photographer Biljana Arandelovic. Rights reserved

Fig. 1.5 Cara Lazara Street in the historical city center. Photographer Biljana Arandelovic. Rights reserved

Fig. 1.6 Kralja Petra Street in the historical city center. Photographer Biljana Arandelovic. Rights reserved

Fig. 1.7 Gračanička Street in the historical city center. Photographer Biljana Arandelovic. Rights reserved

Fig. 1.8 The Bezistan Arcade connects Nikola Pašić Square with Terazije Square. Photographer Biljana Arandelovic. Rights reserved

Fig. 1.9 The National Theatre. Photographer Biljana Arandelovic. Rights reserved

during which 350,000 books and many rare manuscripts dating back as far the twelfth century were lost permanently. The fourth bombing of Belgrade in the twentieth century was by the Allies in 1944, on Easter Day. The fifth bombing lasted 78 days and was by NATO in spring 1999, during which 76 buildings were destroyed and 4,489 damaged. Unfortunately, some of those buildings were part of the country's cultural heritage and listed as national monuments (Perović and Žegarac 2000).

Because of its recent past, Belgrade has a negative global image and is still searching for a new identity by which it will be recognized. Serbia has often been condemned by the international community as the main culprit for the collapse of the former Yugoslavia and the civil wars in Croatia and Bosnia. This has certainly affected its reputation in the world, with Belgrade as the capital city and symbol of Serbia. In some studies of the image of Serbia, the wars and destruction during the 1990s are indicated as the reasons for the negative perception among potential tourists. Security, as an emotional experience related to the destination user, is a serious reason for some choosing to bypass Serbia as a tourist destination (Howard and Allen 2008).

Even though Serbia is still not completely free of the negative political image, visitors to Belgrade have a positive experience of the city. The hospitality of local people is what researchers indicate as the strongest impression on visitors. When asked about their stay in Belgrade, 40.5% of respondents indicated that as the answer to the question about their experience (Armenski et al. 2009). According to some research, the image of Belgrade is developed in the sphere of entertainment

Fig. 1.10 The modern building of Serbia's National library was designed in 1966–1972 by Ivo Kurtović as a freestanding building in the area of St. Sava plateau, with a surface of 30,000 m^2. Although it is a work of contemporary architecture, Kurtović added some elements of a traditional Serbian "house" to it, such as a hipped roof and a porch. The original building of the National Library, with over 500,000 books and manuscripts, which was located in the city center, in Kosančićev Venac, was burned to the ground in the fascist bombing of 1941. Photographer Biljana Arandelovic. Rights reserved

and nightlife, good food, and the hospitality of the local population. Also, studies have shown that it is necessary to create a unique tourist attraction in order to significantly increase the number of tourists. Foreign tourists have cited two key words as their first association with Belgrade—openness and security—which indicates a positive change of image. Tourists say that Belgrade is fast, wild, and alive (Marković 2017).

Belgrade is undisputedly and by all criteria entering the circle of important and attractive regional and European capitals. This is a city that is rapidly developing and moving away from the years of crisis for which it received a negative image in the world. Belgrade now has an opportunity that it cannot miss. It is already the unofficial capital of Southeast Europe.

1.2 Development of Belgrade to 1914

The first data on the looks, organization, and position of Belgrade's streets and roads come from the Roman period. The Romans conquered Singidunum (the ancient city of Belgrade) by the end of the first century BC and joined it with their province of Moesia. The settlement first formed near the castrum (military camp) and below it, and it developed along what is the present-day King Alexander Boulevard. The center of Singidunum was situated near the present-day Cathedral (Saborna Crkva). One of the most important town roads was the one leading to Byzantium (now Istanbul), where some major public buildings were erected. The road to Byzantium started from the old port, located near the confluence of the Sava and the Danube, went along the Danube slope, partly through what is today Jevrejska Street, next to the present National Theatre building (Fig. 1.9) and the Federal Houses of Parliament (Fig. 1.11), along King Alexander Boulevard toward Mirijevo, all the way to Grocka.

Fig. 1.11 Federal Houses of Parliament. Photographer Biljana Arandelovic. Rights reserved

At the beginning of the fourth century, the Huns destroyed ancient Singidunum in 441, along with other cities in this area. By the beginning of the sixth century, Belgrade was a part of the Byzantine state, and it lay in ruins. In the second half of the sixth century, the first waves of the Slavs appeared. With the coming of the Slavs, the ethnic composition of Singidunum significantly changed. Nothing is known about this settlement until 1878, when it reappears under the Slavic name of Beograd (literally—White City). As the Constantinople church influence prevailed over the Romans in these areas, Belgrade found itself in the sphere of Byzantium and its culture. Byzantium ruled Belgrade until the end of the seventh century, and it was ruled by Hungarians from then on.

Belgrade fell under the control of Serbian rulers in 1284, when king Dragutin inherited it through hereditary rights, reaching its peak in medieval times under the rule of Despot Stefan Lazarevic. In 1403, a vassal agreement between Hungary and Serbia was made, according to which the Despot was given Belgrade. He constructed an almost entirely new fort (Figs. 1.12, 1.13, 1.14, 1.15, and 1.16), which was preserved, with small repairs, until 1690 when a large part of it was destroyed in an explosion. After the Despot's death in 1427, Belgrade once again fell under Hungarian rule. From 1459, the Turks were constantly close to Belgrade. In such circumstances, Belgrade could not prosper, nor could it expand.

When the Turks occupied Belgrade in 1521, the settlement retained its earlier character of a military fort and remained such until 1541. It is considered that present-day Dušanova Street is the city's oldest preserved street, with continuity from the Middle Ages. Theoretically, it is the only preserved street from Roman

Fig. 1.12 Kalemegdan Fortress. Photographer Biljana Arandelovic. Rights reserved

Fig. 1.13 Kalemegdan Fortress. Photographer Biljana Arandelovic. Rights reserved. The Roman castrum (military camp) was built at this location in the first century at the end of the Kalemegdan crest, between the current Victor monument, the Military Geographical Institute, and Zidan Gate, leading to Ružica Church

Singidunum. The period from the fall of Buda to the Ottoman Empire and inclusion of Belgrade in the domain of Buda Province to the making of the first Turkish cadaster inventory of the province in 1650 was the most important period in the urban history of oriental Belgrade (Fig. 1.17). The period from 1560 to 1582 is marked by further developments, above all by an increase in the number of town quarters, and also by economic growth. In this period, the Turkish Empire reached its peak and then also began to decline.

The town was horizontally divided into a fortification (in the north) and suburbs (in the south). The suburbs were further vertically divided into the Sava quarter (in the west), the middle section (on the Belgrade Ridge), and the Danube section (in the east). The general characteristic of the street network in the Sava and Danube quarter was the way it was adapted to the terrain's configuration.

The year of 1688 was a turning point for Belgrade's development under the Turks. The increasing urban development was interrupted, and it would never regain impetus. The period 1717–1739 was the only substantial timeline during the Turkish period (1521–1867) in which the city was not ruled by the Turks. Even though the Austrians had great ambitions to transform Belgrade from an oriental town into a European city, it was too short an interval, so only some of their plans were realized.

Fig. 1.14 Kalemegdan Fortress. View of the Victor monument, one of Belgrade's landmarks, by renowned Yugoslavian sculptor Ivan Meštrović to commemorate the victory over the Ottoman and Austro-Hungarian Empires during the Balkan Wars and the First World War. Photographer Biljana Arandelovic. Rights reserved

Fig. 1.15 Kalemegdan Fortress. Photographer Biljana Arandelovic. Rights reserved

Fig. 1.16 Kalemegdan Fortress. Photographer Biljana Arandelovic. Rights reserved

Fig. 1.17 Street in Dorćol quarter. Photographer Biljana Arandelovic. Rights reserved. Dorćol (dort-jol), a well-known Belgrade quarter, means "four roads" or "crossroads" in Turkish. This title marks the spot where two prominent streets cross at right angles to each other. In the curved organic plan of the old Turkish town, the only orthogonal crossroad was a venue of wonder. A careful examination of successive plans of Turkish Belgrade, made in different time frames through the centuries, shows us that this extraordinarily orthogonal crossroad was maintained practically all the time

According to the Treaty of Belgrade of 1739, Belgrade was given back to Turkey. In this period, Belgrade once again became a border fort. Typical of that period was the return of the oriental look to the urban structure of the town, and the renovation of old mosques, inns, public baths (hammams), and other public and residential buildings. However, there was no territorial expansion.

The beginning of the nineteenth century was a time of great national uprisings. The first Serbian uprising under Karadjordje broke out in 1804 (Meriage 1978), but the insurgents only managed to occupy Belgrade in late 1806. During the preparations for a siege one night, Savamala, an earlier Lower Serbian town, was set on fire. New Savamala was founded closer to the Ditch, around Karadjordjeva Street.

Belgrade did not remain in Serbian hands for long, and Serbia, according to the Treaty of Bucharest in 1813, returned to Turkish rule. The favorable international situation made a way for the Second Serbian uprising.

Immediately upon the city gaining autonomy, Belgrade's history saw great changes in the nineteenth century. Within a span of only several decades, Belgrade significantly changed its appearance: from being a derelict, poorly arranged small town with winding, narrow streets, it started to become the principal economic, cultural, and governmental center of the young Serbian state, through its planned reconstruction. From 1815 until the 1880s, Belgrade was surrounded by the Ditch and palisades. The entrance to the city was through four gates, the most important being the Stambol gate, in the location of present-day Republic Square, from where the road to Constantinople began. The town was divided into small, twisting, and narrow streets, with poor pavements and no street lamps (Djurić-Zamolo 1977).

Serbia gained its autonomy in 1830 and Belgrade became the capital of the Principality of Serbia. After 1830, when Knez (Prince) Miloš was pronounced hereditary prince of Serbia, the influx of the Serbian population from all over Serbia, Vojvodina, and other regions still under Turkish rule increased. After 1829, Prince Miloš ordered the construction of a number of important public buildings, and in this way formed the Serbian center within the city. Finally, Belgrade officially became the capital of the Kingdom of Serbia in 1883. All of Serbia's foreign trade went through Belgrade, as it was the only railway connection to Europe, particularly to Austro-Hungary. Such circumstances were favorable for its development. Apart from being the economic and transportation center, Belgrade also became the center of the government apparatus, and of the political and social life and main cultural and educational events. Belgrade accumulated most of the capital in Serbia (Maksimović 1983).

The Serbs' wish was to erase everything reminding them of the Turkish period. They turned to the West and its culture, which made a great contribution to the cultural transformation of the Serbs in various fields of human activity. Radical advance in the construction and development of Belgrade was made by Emilijan Josimović, who recorded Turkish, oriental Belgrade in the Ditch, and then provided a proposal for the regulation of this space in 1867.

The attractive power of Belgrade was obvious, seen in the increase of its number of citizens, especially after 1887, when the first railway line passed through Belgrade. Belgrade had 35,483 citizens in 1884. After 16 years, this number doubled to 69,769 and in 1910 its 89,876 citizens brought it close to the size of large European cities (Maksimović 1983).

By the end of the nineteenth century, handicrafts were the most significant branch of the economy. Undeveloped agriculture and small-scale industry and trading thwarted the economic development of the country. Only after full independence in 1878 did the first industrial manufacturing appear. Through the legal acts of 1873 and 1898, special preferences were given to industrial contractors. On the eve of World War I, in 1913, Belgrade and Zemun had 55 companies with around 5,000 employees, and more than 100,000 citizens (Petrović 2001).

Austrian urban heritage from the first half of the eighteenth century is clearly visible on structures in the city center, from the time when Belgrade belonged temporarily to the European circle of late Baroque and began renovation following the influence of the Habsburg capital, Vienna. This can be seen in the examples of neoclassical facades in the center of Belgrade. Belgrade is rich in buildings which were directly influenced by Viennese Secession and built at the beginning of the twentieth century. These buildings were designed by Serbian architects who were educated in Austria and later came back home (Samardžić 2015). Many of these architectural examples are preserved as cultural monuments (Figs. 1.18, 1.19, 1.20, 1.21, 1.22, and 1.23).

The Austro-Hungarian Empire was located in the close vicinity of Belgrade, on the other side of the Sava and Danube, which, until the early twentieth century, meant that it was a catalyst for the modernization and Europeanization of Serbia. Achieving formal independence was particularly felt in Belgrade, as a result of which the improvement of relations with Austria-Hungary Empire was felt: railway construction, navigation regulation on the Danube, and creation of new economic ties with other Balkan countries (Arandelovic et al. 2017).

1.3 Belgrade Between Two World Wars

In the aftermath of WWI, in 1918, Belgrade became the capital of the newly formed Kingdom of Serbs, Croats, and Slovenes. This unity was formed by the Serbian Karadjordjević Dynasty under King Petar I, whose primary aim was to unite all South Slavs in one country. The nation of Yugoslavia was born. The Kingdom of Serbs, Croats, and Slovenes changed its name in 1929 to the Kingdom of Yugoslavia. The beginning of WWII was the end of the Serbian Karadjordjević Dynasty and the end of the Kingdom of Yugoslavia (Šuvaković 2003).

The Kingdom of Yugoslavia was heavily influenced by having a dominant traditionalist monarchy, the Karadjordjevic Dynasty. Modernism was aimed at integrating the Kingdom with its international contemporaries, defining the transition from a rural country to a modern, industrial system (Kulić 2009). Yet, this goal was not completely achieved due to a large number of opponents of the Serbian Dynasty's leadership emerging throughout the whole country. They strongly opposed the modernization envisioned by King Petar I Karadjordjević.

After the end of WWI, the significant influence of Paris was noticed with regard to new visions and demands for the development of cultural life. Paris was traditionally attractive to Serbian intellectuals and elite, but war circumstances reinforced this. After withdrawing through Albania (Figs. 1.24 and 1.25), thousands of Serbian refugees went to France, and most returned to Serbia bringing back the Parisian artistic spirit. After the war, around 8,000 students came back to Serbia. Many of them took leading positions in the government and were among the social elite. This increased French influence in politics, science, culture, and education (Dimić 1997).

Belgrade, as the capital of the newly formed Kingdom of Yugoslavia, was trying to assume a position equal to other European capitals, by changing, modernizing,

Fig. 1.18 Athens (Atina) Palace was built in 1902 in academic style by renowned Serbian architect Dimitrije Leko as a residential and commercial building for the merchant Djordjo Vučo. Rich in ornaments and color, the building has had a prominent position at Terazije Square 28 ever since. Photographer Biljana Arandelovic. Rights reserved

INGLOT
INGLOT

◂**Fig. 1.19** Merchant Stamenković's House, Kralja Petra Street 41. Photographer Biljana Arandelovic. Rights reserved. This cultural monument, built in 1907 by Andra Stevanović and Nikola Nestorović, belongs to a group of typical residential buildings built in Belgrade before the First World War. The architects achieved harmony between academic concepts and art nouveau decorations. The vertical division of the main facade was achieved with characteristically pronounced pilasters, while the open facade's surfaces are covered with green ceramic tiles, because of which the building is popularly called the "building with green tiles"

and accepting European models of life. Under these circumstances, it was some eminent individuals, as well as institutions, that played a crucial role in achieving this goal. One of them was the Association of Friends of the Arts Cvijeta Zuzorić (Figs. 1.26 and 1.27). Almost without exception, all of the most prominent personalities from the cultural world at that time took part in this association and its work.

Artistic life in Belgrade flourished in the period between the two World Wars, and it became Belgrade's most productive period in terms of art and culture. Artists were open to new aesthetics and were looking toward Europe for inspiration. In this sense, the Association of Friends of Arts helped Belgrade audiences grasp the broader context of European art. The association also played a serious role in the process of women's emancipation in Belgrade by involving them in organizing committees for cultural events, thus making an important contribution to the general modernization and Europeanization of Belgrade between the two World Wars (Vučetić-Mladenović 2003).

In terms of architecture, new generations of young people traveling and being educated in Western European cities made it possible to keep in touch with new tendencies and fashions. Many architects had a direct opportunity to learn from the pioneers of modernism and to bring back new ideas (Fig. 1.27). The result of this was the formation of the Group of Architects of the Modern Movement (GAMP) in Belgrade in 1928 by a group of young modernists, who in the following years shaped the architectural scene. The house of one of the founders of the GAMP group, Zloković House, by Milan Zloković is considered to be the first modern house in Belgrade (built in 1929) (Blagojević 2003).

Serbia, which was located between the Western world for which it had long craved and the residual Ottoman Empire, still managed to develop a significant number of modernist buildings, some of which can still be found in Belgrade (Blagojević 2003).

The process of economic development was slow in Belgrade after WWI. During the first 10 years, no new bigger or modern factories appeared. The renewal of the industry that had been destroyed had started by 1925, but the amount of production from the pre-war period was never again reached. In 1919, Belgrade had 67 industrial corporations, in 1925 there were already 99, and 2 years later, the city boasted 138 corporations (Piljević 2001).

ДОМ
ВУКОВЕ
ЗАДУЖБИНЕ

◂**Fig. 1.20** The Ministry of Education building was built in 1870–1871 following architect Aleksandar Bugarski's project; it is located in Kralja Milana Street 2. The main facade was transformed into Serbian-Byzantine style in 1912 by architect Branko Tanazević. The base of the building retained the old scheme, but the elements of the Serbian-Byzantine style of the main facade, combined with Art Nouveau, are new for the period in which they occurred. The red ceramic decoration in the style of the Morava school, and the coat of arms of the Kingdom of Serbia are well embedded in this Art Nouveau facade. Photographer Biljana Arandelovic. Rights reserved

Work on the new infrastructure, as well as construction, began several years after WWI ended and most improvements were carried out on public utilities. The street network was reconstructed to adapt it to the new traffic requirements, as most of the paving was inherited from Turkish times. A large number of new green markets were opened, and electricity was introduced in many households. For the first time in the history of the city, road bridges were constructed across both rivers, the Danube and Sava (Maksimović 1983).

The greatest number of buildings was reached in 1932, when collective housing construction increased abruptly. According to the first census after World War I, in 1921, Belgrade had 7,465 houses (300 built after the war) with almost 28,000 housing units (Piljević 2001). By the end of 1938, more houses had been built than Belgrade had had before the war. Whoever had money invested it in a house, which led to negative economic consequences. The newcomers to the city, whose number grew, did not lose their contact with the countryside. This explains how poor inhabitants managed to survive the period of crisis between the two World Wars (Marković 1992).

The influx of poor immigrants, who built small individual houses, contributed to the fact that Belgrade expanded spatially and swallowed up suburban districts (Pašino Brdo, Zvezdara, Rakovica, Voždovac). In addition, villa quarters for the wealthy population were built (Dedinje, Topčidersko Brdo, Kotež Neimar). In 1921, Belgrade had 135,000 inhabitants, in 1931 that number reached 266,000, and in 1940 as many as 320,000 inhabitants. Belgrade's territory on the eve of WWII was three times larger than in 1918 (Maksimović 1983).

The building area of Belgrade expanded rapidly, following demographic growth. Developers bought farming land, divided it into construction lots, and sold it to potential builders. The city tissue was being thoroughly reconstructed and signs of the oriental town vanished almost completely. The height of the buildings increased, and the general conditions of housing and living in the city improved significantly (Bojović 2003). At the same time, many public and commercial buildings were built, for example, the National Assembly House, and also the Ministry buildings in Nemanjina Street (Fig. 1.28), St. Mark's Church (Fig. 1.29), the State Printing Office BIGZ, and the Air Force Command in Zemun.

Fig. 1.21 Moskva Hotel (Hotel Moscow) is the first building that marked the future construction of Terazije Square. Hotel Moscow, built in 1907, is among the first examples of Art Nouveau style in Serbia. The building is one of the recognizable symbols of Belgrade. Photographer Biljana Arandelovic. Rights reserved

Fig. 1.22 Hotel Moscow. Photographer Biljana Arandelovic. Rights reserved

Fig. 1.23 King Peter I elementary school, completed in 1906 and located in Kralja Petra Street 7, in the old city center. Due to the different arrangements of the rooms and the great attention given to hygiene facilities, the school was the most modern school building in Serbia at the time. It was designed by Jelisaveta Načić, remembered as the first female architect in Serbia. Photographer Biljana Arandelovic. Rights reserved

КОМЕРЦИЈАЛНА БАНКА
MaxMara
MaxMara

◂**Fig. 1.24** Albania Palace is located at the beginning of the main town axis that extends from the central commercial space on the top of Terazije to the highest ground level of Vračar hill, where St. Sava Temple (Fig. 1.30) is located. Built in 1940 after two successive competitions in 1938, the palace presents the spatial urban emphasis of Terazije and is one of the most recognizable symbols of Belgrade. Branko Bon and Milan Grakalić won the first competition, while Miladin Prljević won the second. His project adopted much of the winning solution from the previous competition. The palace was named after the inn which had been previously situated there, and which got its name in memory of Golgotha of the Serbs while passing through Albania in World War I. This structure represents a big achievement for the Serbian architecture of the time. Albania Palace was designed in the spirit of the international modernism of the 1930s, with marble facades, and for a long time it was the tallest building in Belgrade. The building is defined by its slightly convex shape, an open ground floor with shops, and its dominant vertical high volume. Photographer Biljana Arandelovic. Rights reserved

Despite these positive developments, fairly uncontrolled territorial development of the city continued, so just before WWII, a decision was made to make a new general plan, which was eventually completed in 1939 (more in the chapter Belgrade's urban plans through history). The most important achievement in this phase of Belgrade's development is the fact that it did become a European city; and it was built according to the Building Law of 1931, which was the first real European law in Serbia dealing with town planning. The period between the two World Wars is known as the Golden Age of civil Belgrade (Figs. 1.24, 1.25, 1.31, 1.32, 1.33, and 1.34).

1.4 Belgrade from the End of WWII Until 1991

After WWII, significant political and social changes took place. Socialist Yugoslavia was formed from the remains of the Kingdom of Yugoslavia in the aftermath of WWII. Belgrade suffered massive devastation during WWII, when around 40% of all the building capacities were either destroyed or damaged (Bojović 2003). Yugoslav propaganda was oriented toward Moscow, and Belgrade became the seat of the Cominform (the Information Bureau of European communist parties) and the first international communist forum after the dissolution of the Third International, in 1947 (Kulić 2014a, b).

After Tito (Fig. 1.35) rejected making Yugoslavia dependent on the Soviet Union, Yugoslavia left the Soviet bloc in 1948 and changed its political course toward the West. The country was close to joining NATO in 1955, just before collaborating again with the Soviet Union and establishing a new position between the two blocks. After Stalin's death, Yugoslavia gained a special status in the communist bloc, thanks to its socialist revolution from the recent past. Yugoslavia, unlike other European countries where the communist upheaval came with the arrival of the Red Army, had a predominantly indigenous communist movement and therefore was considered the land of authentic socialism (Zec and Radonjić 2012).

Fig. 1.25 Albania Palace. Photographer Biljana Arandelovic. Rights reserved

Fig. 1.26 Cvijeta Zuzorić Art Pavilion. Photographer Biljana Arandelovic. Rights reserved

Fig. 1.27 Cvijeta Zuzorić Art Pavilion. Photographer Biljana Arandelovic. Rights reserved

Fig. 1.28 The Palace of the Ministry of Finance of the Kingdom of Yugoslavia (today the building of the Government of the Republic of Serbia) is an anthological work of monumental Serbian interwar architecture and the most important achievement of Russian architect Nikolaj Krasnov. It was built in 1926–1928, with an extension to the third floor built in 1938, at Kneza Miloša 22. Krasnov was a key figure in the architectural development of Belgrade between the two wars; he came to Yugoslavia as an immigrant and later became the head of the Department of Monumental Architectural Developments and Monuments, which was part of the Serbian Ministry of Housing and Building. Photographer Biljana Arandelovic. Rights reserved

There are two key periods in the post-war era crucial for understanding the economic situation. First is the period of the centralized bureaucratic system, implemented as a copy of Soviet experience, that lasted while relations with the Soviets were economically ideologically, politically, and militarily close (1945–1950). The second period is marked by the construction period of the self-management system (1950–1988), which marked the break with the Soviet Union, worsening political relations almost to the brink of armed conflict; approaching the West; and as a grand finale the establishment of the Non-Aligned Movement, which positioned Yugoslavia in the major global political arena (Zec and Radonjić 2012).

In this way, Yugoslavia participated in all sides of the Cold War. The country positioned itself in a particular way between the two blocks of the Cold War, softening the contrast between socialism and capitalism. It became part of a Third World of developing countries pursuing decolonization and modernization, in contrast to the First World of developed Western liberal democracies and the Second World of socialism (Kulić 2014a, b).

50% of the industrial capacity was destroyed in WWII. The intensive development of energetics, electrical, chemical, and mechanical engineering and the nutrition industry appeared after 1960. Nationalization and confiscation of industrial corporations occurred in the aftermath of WWII (in the pre-war period property was private), and the post-war period was marked by public property. There were just 118 industrial corporations in 1950, as compared with 230 in 1940. Numerous former small metal factories disappeared as a result of industrial multiplication, while small-scale old handicraft facilities along Knez Mihailova Street also vanished (Petrović 2006). During the 1960s, the deficit in the trade balance was covered by foreign aid and war reparations.

The region of Belgrade covered 1.3% of Yugoslav territory and 6.5% of the country's population in 1985. The region had 6.3% of all employees in the Yugoslav economy, and 6.4% of the national income. 1,582,000 employees worked in the Belgrade region, or 102 employees per 1000 inhabitants (Grčić 1993).

A new state model of development was based on accelerated industrialization, but in the early 1960s the number of factories in Belgrade did not increase. Many factories moved to other parts of the country such as the northwestern and central parts of Yugoslavia, for political reasons. Belgrade was planned to be developed as the new political and administrative capital, and the move to transfer industry to other parts of Yugoslavia was logical. In 1961, the metal industry, with its 27 factories, was Belgrade's most important branch. In the same year, the statistics showed that around 30% of employees in Serbia were employed in Belgrade's metal industry (Petrović 2006).

Fig. 1.29 St. Mark's Church. Photographer Biljana Arandelovic. Rights reserved

The majority of newcomers came from the villages to the city as a consequence of the informal policy of the communist government, which was to develop industry and destroy agriculture. Significant profit was spent on incomes and social benefits for newcomers, while long-term investment in production was neglected.

Migration to Belgrade was prompted not only by extensive prospects of jobs for unqualified manpower, but also by systems of selective social (pension and health) insurance as one of the characteristics of the communist system. Insurance was a special privilege of employees in public enterprises and their families, and people used it to its maximum. The new social security system did not much care for developing villages, which forced farmers to look for better opportunities by moving to Belgrade, where the offer for employment in public enterprises was always open. Additionally, with the employment of at least one member of the family in a factory or another public enterprise, the whole family received benefits. This system supported and motivated workers (Grčić 1993).

Fig. 1.30 Saint Sava Temple (Hram Svetog Save) is located on the Vračar plateau in Belgrade's Vračar quarter. It is considered to be one of the largest Orthodox churches in the world. The monumental white marble and granite facade is topped with a copper dome, while the interior is still incomplete. St. Sava Temple is another eternal work in progress, whose construction has lasted for more than 100 years, since the initial plan was made in 1893. The Temple is dedicated to Serbia's first archbishop. Photographer Biljana Arandelovic. Rights reserved

Simultaneously another problem developed that of informal construction, which would later become a major problem in Belgrade. It became a phenomenon in the middle of the twentieth century, either as a tolerated or unnoticed addition to contemporary urban development. Furthermore, in the period of socialism, informal construction intensified, starting in the early 50s and advancing into the 70s (Topalović 2011).

International credit and loans taken out to improve the national economy were not used for their basic purpose, which resulted in the crisis in the 1980s, when vast resources were spent on paying back huge loans. Some corporations were subsidized from the state budget, sustained as part of the economic market of the country, although they produced only losses.

This period could be depicted as the self-governing paradigm when companies managed by workers made decisions about the distribution of income. This formed the foundations of a new, authentic self-management system that became a basic feature of the socioeconomic model until its collapse in the 1990s. The economic foundation of this model was public property as a form of ownership that was not present in any other country. The inefficient economic system led to a large discrepancy between the reported spending and lack of financial resources to support it. As a result, the system deficit over four decades was fictitiously covered in different ways. The financial system destabilized Yugoslavia, and the centralized party structure was the only element that held it together. The main role of the huge administration was to redistribute funds. The state controlled 70% of all investments. When Tito died, who was the last binding element and indisputable authority in Yugoslavia, the collapse could not be avoided (Zec and Radonjić 2012).

The transition to market liberalism took place in the period 1963–1973. The greatest misconception was that companies based on social ownership in competitive circumstances would behave like a privately owned company. During this period, a rapid growth in household incomes was reported, but economic growth had stopped and the situation worsened, with inflation and unemployment. Self-government was not working. What followed was an inflation rate above the interest rate and the capital flow between the Republics vanishing. In the period 1971–1980, the national income grew at an average annual rate of 4.8% while at the same time net foreign debt, on average, annually, grew nearly four times faster —18.6% (Zec and Radonjić 2012).

This was the beginning of Yugoslav disintegration and the creation of the individual economies of each Republic. The power of the Communist Party was reduced in the midst of economic crisis. Economic growth in Yugoslavia was fictitious. The next step was logical; the process of federalizing the party began and thus became a platform for the future splitting of the country. The abolition of social ownership didn't occur until the 1990s, when it was already too late. The League of Communists of Yugoslavia broke up, and national parties within the Republic were formed, which was the beginning of the creation of new states.

Fig. 1.31 Residential buildings in Vlajkovićeva Street. Photographer Biljana Arandelovic. Rights reserved

Fig. 1.32 The Aero Club building at the corner of Uzun Mirkova Street and Kralja Petra Street was built by Vojin Simeonović in 1934–1935 for the National aviation facility of The Kingdom of Yugoslavia, in art deco style. Photographer Biljana Arandelovic. Rights reserved

Through this period, Belgrade passed through various ups and downs. The period from 1960 to 1976 is underlined as the period of the most intensive and comprehensive development of Belgrade in 200 years. It was based on the general city plan of 1950. Financing was provided by massive housing funds and city institutions. The basic concept of development was realized on a massive scale on the mostly empty land. Increased housing construction on the fringes of the city territory was a response to an immense immigration wave. The investment effort

Fig. 1.33 The Patriarchy building, Kneza Sime Marković 6. Photographer Biljana Arandelovic. Rights reserved. The Patriarchy building was built in 1935 by Russian architect Viktor Lukomski in a modernized Serbian-Byzantine style. Immigrant Russian architects built Belgrade's monumental government buildings and the court between the two world wars, and made a big contribution to the architectural development of the city. Lukomski was one of them

Fig. 1.34 Igumanova Palace, Terazije Square, 1936. Photographer Biljana Arandelovic. Rights reserved. The Krstić brothers won first prize in a 1935 competition for the realization of the office building for the Sime Igumanova Fund (Igumanova was a great Serbian benefactor of the nineteenth century). This town palace is one of the most important works of Petar and Branko Krstić (two of the most important Serbian architects in the twentieth century), and also one of the key buildings reflecting the process of transforming the traditional academic concept into the modern

Fig. 1.35 The May 25 Museum was designed by Mihailo Janković and built in 1962 as a gift from the City of Belgrade to Josip Broz Tito on his 70th birthday, with the purpose of hosting all the gifts Tito received during his leadership. The museum has a symmetrically arranged rectangular form, with a butterfly like roof, massive pillars on the ground floor and a symbolic mosaic above the main entrance to the building. The complex consists of a wide prospectus with a cascading processed plateau, fountains, benches, a small amphitheater, and green space. The idea of admiration for the authoritarian personality is embedded in the entire architectural and urban concept. The synthesis of architecture, sculpture, and painting is an expression of the time in which the museum was created. The museum is one of the few buildings with completely preserved original features and architecture from the period of socialist Yugoslavia. It attracts large numbers of visitors, especially those from the former Yugoslavian republics. The 25 May Museum is located in the elite Dedinje district of Belgrade. Photographer Biljana Arandelovic. Rights reserved

entirely transformed Belgrade. New Belgrade, parts of Zemun, Banovo Brdo, Konjarnik, Šumice, and Karaburma were intensively built. These parts of the city had previously almost no inhabitants. It is important to mention suburban villages, the appearance of which changed completely in this period (Bojović 2003). In 1947, suburban settlements were still small. Only Železnik, Batajnica, and Žarkovo had between 4,000 and 7,000 inhabitants. Most suburban settlements were similar to villages (Petrović 2006). Intensive development gave Belgrade a new form, and at the same time its organization and way of functioning changed. However, the existing parts of the town, and especially its core, were neglected. The city center did not manage to penetrate the suburbs in terms of spatial–functional symbols, or in terms of social and cultural influences. As Belgrade spread, there emerged

Table 1.1 Population of Belgrade after WWII

Year	1945	1951	1961	1981
Population	313,000	428,000	843,000	1,470,073

Source Arandelovic et al. (2017)

multiple concentric circles of new suburbs, filled with mass residential buildings. Political and cultural authorities did nothing to help the social and cultural development of newcomers through various activity programs. Suburban zones remained undeveloped from the time of their original construction (Table 1.1).

The end of this period brought unexpected events. In 1989, the same year that the Berlin Wall came down, Slobodan Milošević was elected president in Serbia. He was not interested in maintaining positive relationships with Third World leaders, and in 1992 Yugoslavia was expelled from the Non-Aligned Movement. At the same time, the war in Yugoslavia continued.

1.5 Belgrade and the Non-aligned Movement

The Non-Aligned Movement was inaugurated in Belgrade in 1961. The Conference of the Heads of State or Governments of Non-Aligned Countries gathered the leaders of 25 developing countries and approximately 40 delegations of diverse anti-colonial liberation movements. The inauguration event took place in Belgrade, September 1–4, 1961.

Tito's leading role in creating the Non-Aligned Movement was the result of Yugoslavia's ambitious vision to find its position among superpowers rather than merely maintaining independence. The Non-Aligned Movement was not only an organization, it was more an ideology based on the renunciation of the premise that developing countries must follow the ideology of one of the superpowers in the Cold War conflict. At the time when the Berlin Wall came down, the Non-Aligned Movement had more than a hundred member states. Belgrade also benefited from student and expert exchanges with African countries, India, and the Middle East (Jakovina 2011; Willets 1978).

As the Non-Aligned Movement member states created their own global network of cultural and economic exchange, Yugoslavia fully used this opportunity. The result was that Yugoslav architects found commissions in the Middle East, Africa, and Latin America. Yugoslav construction companies had large-scale operations worldwide. As the capital of Yugoslavia and cofounder of the Non-Aligned Movement, Belgrade was in the privileged position of being able to export its expertise to developing countries in the movement during the 1970s and 1980s.

> Tito's Non-Aligned Movement reached places where American, English, and Russian superpower influence could not easily go: the Middle East, Africa, and Latin America. But how could high-caliber practitioners like Charles Jencks or Kenneth Frampton – not to mention Bruno Zevi or the utopian Doxiades – ever swallow Yugoslav architectural

experiments in Libya with Gaddafi; in Iraq with the early Saddam Hussein regime; and architectural manifestations of suspect business practices in Zimbabwe under the cruel dictator Robert Mugabe? (Weiss 2013: 95)

One of the biggest domestic beneficiaries of the contracts between non-aligned countries was Belgrade design and construction corporation Energoprojekt, which worked on projects in more than 80 countries on four continents (architectural, urban, and infrastructure projects). Weiss (2013: 95) described Energoprojekt as "the socialist version of a Skidmore, Owings and Merrill-type corporate architecture firm," saying that "these designs combined late-modernist corporate style with an oriental influence that allowed state companies to appear local in the context of Middle Eastern and African commissions."

The state promoted architectural and engineering firms as well as engineers and architects as a brand to the Non-Aligned Movement countries, which resulted in productive cooperation between Yugoslavia and Non-Aligned Movement countries. Tito used "soft socialist power" to establish commissions and boost the construction industry. One of the most successful architectural teams was Dragoljub and Ljiljana Bakić, the husband-and-wife team who worked independently in their own company, called Atelje 5, within the Energoprojekt Company, and whose structures were built on all four continents, but mostly in the Middle East and Africa. They also created some important constructions in Belgrade, such as Pionir Hall and Višnjička Banja, a modern planned settlement in the east part of Belgrade. Architect Dragoljub Bakić is today involved in civic activism in Belgrade and is one of the biggest critics of the current city government and the way it approaches urban development and reconstruction in Belgrade.

1.6 Belgrade from 1991 to 2000

During the year 1989, economic crises culminated and the economy entered a zone of hyperinflation in which the general level of prices in the period December 1988–December 1989 increased by 2714% (Zec and Radonjić 2012). This was the second highest hyperinflation in world economic history. The breakup of socialist Yugoslavia was partly due to the inability of political elites to radically transform the economic and political system, although the collapse of the Yugoslav federation in the 1990s was a result of complex internal and external geostrategic circumstances, culminating in violent political conflict. The breakup of the federation was expected as the end of a process which had already lasted a long time. The situation in which the country found itself was extremely difficult.

Yugoslavia's disintegration led to the erosion of its single market. Yugoslavia's disintegration was violent, including civil war on the territory of the former Republics of Bosnia and Herzegovina and Croatia. The Yugoslav separation process surprised everyone, not only because of its suddenness, but also because of its brutal nature (Arandelovic et al. 2017). The United Nations imposed sanctions on

Serbia. One outcome of the wars in the 1990s was that Serbian refugees flooded into Serbia, many of whom came to Belgrade. The '90s ended with the NATO bombing of 1999. The last decade of the twentieth century will be remembered as the period of the self-destruction of Belgrade. After disintegration of the Socialist Federal Republic of Yugoslavia, the new Federal Republic of Yugoslavia was formed out of two republics, Serbia and Montenegro, in 1992.

The Serbian Socialist Party was built from the Serbian Communist Party, with Slobodan Milošević as its leader. This was the ruling party throughout the 1990s, until the elections in 2000. Nationalist ideology prevailed in the second half of the 80s and the communist elite continued to manage post-communist Belgrade. Because of its strong political, economic, and cultural impact, the fall of communism in Belgrade lasted for 10 years, much longer than in Eastern Europe, where communism was a product of Soviet occupation (Arandelovic et al. 2017).

The new Federal Republic of Yugoslavia saw the shift toward a parliamentary political regime but very few serious reforms were initiated. The privatization of socially owned companies occurred very slowly, while the most important companies were nationalized by members of the ruling parties, and party members took leading positions in them (Petrović 2001).

A program of economic recovery was initiated at the beginning of 1994, after hyperinflation in 1992 and 1993. Because of hyperinflation, employees and employers have been extremely motivated to buy shares, and as a result, 43% of the total capital of Serbian companies was transformed by mid-1994 (Hadžić 2002).

Despite this, market-oriented reforms were not continued and as a result, economic stabilization and growth were not constant. Unfortunately, the Federal Republic of Yugoslavia was ranked in the third group of economies in transition regarding its achievement of reforms. After the Dayton agreement in 1995 and the lifting of sanctions in 1996 and 1997, international conditions for privatization were slightly better.

Belgrade's population increased drastically during the 1990s. Belgrade, as the capital of the Federal Republic of Yugoslavia, had 1,602,226 inhabitants in 1991, after which nearly 600,000 people arrived as refugees from Croatia and Bosnia and Herzegovina (Petrović 2001). An outbreak in the number of illegally constructed housing came as the aftermath of these traumatic circumstances.

Not only the ongoing political crisis, but also the enhanced feeling of a loss of perspective for a large part of the population, in particular, for the young generation, were the main factors resulting in intense emigration during the 1990s (Lukić and Anđelković-Stoilković 2017). The main reason behind the new wave of emigration was the hard economic and political situation forcing thousands of young intellectuals to leave the country for the possibility of a better future.

The social structure of Belgrade changed. At least several tens of thousands of the most educated members of society fled, while Belgrade was receiving the majority of the refugees, mainly from rural parts of the former Yugoslavia affected by the war. Some parallels with the situation from the beginning of the twentieth century are obvious from when France was the popular destination for young Serbian intellectuals. However, this situation was very different, and the 1990s brought on a new wave of emigration. The difference from the beginning of the twentieth century, when young intellectuals went to France, Austria, or Germany in order to educate themselves and learn more about European achievements, eager to escape from the Ottoman legacy, and accept the new and the modern, was that the new emigration in the 1990s was to an incomparably greater extent, and the people were full of despair and without much hope. Young people had once gone to Western European countries in order to come back and assist their own country's development. Young and educated people leaving the country at the beginning of the 1990s just wanted to escape from the regime and start a completely new life, without any intention of coming back. The most popular destinations for the new emigrants of 1990s were countries such as the USA, Canada, and Australia, where highly educated young experts could easily find appointments.

The comprehensive decline of municipal powers was the consequence of a full decade of economic crisis. Additionally, the local planning system nearly collapsed (Hirt 2009; Zegarac 1999). Illegal construction, which was inherited from socialism, increased during the 1990s. In addition, during the period of hyperinflation, the streets of Belgrade were swamped by dealers of currencies and gasoline, street vendors, and kiosks, who could be found in unforeseen places. The United Nations embargo (1992–1995) helped distribute informal activity far beyond housing production. This provoked a situation in which many of the city's residents were considered to have some connection with illegal activity or business, such as the gasoline trade, informal public transport, or street trade (Topalović 2011).

If we compare the situation with the illegal construction during the socialist period, clear differences can be observed. During the socialist period, illegal construction was associated with marginalized social groups, while here the players changed to include groups of people in power as well. Another difference is the location of illegal construction. The previous illegal constructions used to appear on the periphery of Belgrade, while the new ones became common in the central city areas, too.

Members of the political elite participated in illegal construction with no fear and no sense of destroying the city's appearance (Hirt 2009). Those new illegal urban structures appeared in luxury residential areas in Belgrade like Dedinje and Senjak, for example. Indeed, entire new blocks, such as Padina, were established in this manner. The new dwellings used a highly lavish and kitschy style, and a mixture of aluminum and reflective glass, giving a high-tech look, in combination with some traditional architectural segments. The name for this phenomenon, "turbo-architecture," came from a new music genre called turbo-folk which emerged in the '90s on the Serbian scene and took its influence to all segments of society Weiss (2007).

1.7 Belgrade at the Beginning of the Twenty-First Century

After the critical 1990s, political changes that appeared in October 2000 opened new perspectives for Serbian society and created a positive overall projection for the future. Serbia elected its first post-communist democratic and reformist government in 2000. During January 2001, the winning coalition of 18 political parties formed the government. A transitional government under the socialist's presidency enabled political and economic transition in the country. The socialist regime renounced its political but not economic power. In the period from October (2000) to January (2001), during privatization, the political wing of the socialists privatized 800 successful state-owned companies (Antić 2019). There is a prolonged history of privatization in Serbia, which was initiated in 1990 and was only fully completed according to privatization legislation in 2015. However, privatization only became serious after the democratic revolution in 2000 (Vujačić and Petrović Vujačić 2016).

The Federal Republic of Yugoslavia returned promptly to the international community and organizations. The country returned to the United Nations and made a series of important steps toward integration into the European Union. The most important steps toward obtaining financial stability were taken, namely, the return to international financial institutions such as the International Monetary Fund and the World Bank. The Federal Republic of Yugoslavia again changed its constitution and name in 2003, when it became the Union of Serbia and Montenegro, which lasted from 2003 to 2006. Finally, in 2006, after Montenegro's secession from the union with Serbia, both former republics became sovereign.

Zoran Djindjić, the first democratically elected prime minister, was assassinated in March 2003. His assassination was executed by members of the Unit for special operations, JSO (Jedinica za specijalne operacije), which was an elite special forces police unit within the State Security Service, but the political background of the murder has never been solved. A number of possible reasons led to Djindjić's assassination. In the years following the political changes in 2000, the democratic government, made up of several parties, was constantly divided. They never had a stable political majority and continuously depended on support from some of the parties from the former regime. This prevented real reforms, reinforced by international requirements, such as the Kosovo issue and cooperation with the Hague war crimes tribunal. Not all parties in the democratic government coalition had the same vision of the future and how to solve problems inherited from the past regime. Neither did they share the same attitude toward entering the European Union. After Slobodan Milošević's deportation to The Hague, discontent among some of the public also strengthened.

Reforms were too slow, as was privatization. The government tried to minimize the negative mood in the nation by increasing pensions and salaries. The estimation is that since the beginning of the transition in Serbia the GDP has grown: cumulative GDP growth in the period 2000–2008 was 47% and was about 5% a year on average. The average GDP growth rate over the entire period 2000–2012 was around 3% a year, which was slightly less than the average of CEE countries in which growth was 3.4% (Arsić 2013).

Studies show that all of the important indicators (social, economic and environmental) have declined, resulting in decisive consequences for urban and regional development.

The general trend can be defined as increase without development, since there has not been a radical restructuring of the economy or society (Zeković et al. 2015).

The next big political change happened in 2008 when the Democratic Party (of Zoran Djindjić) won the election but not decisively enough to govern, and its coalition partner became the Socialist Party (of Slobodan Milošević). In that year, the former right-wing Radical Party was transformed into the Serbian Progressive Party (SNS), the party under the leadership of Aleksandar Vučić (former Prime Minister and current President of Serbia). The Serbian Progressive Party won the 2012 election and has ruled since then by gaining a majority in the elections.

In 2012, Aleksandar Vučić announced his idea for the Belgrade Waterfront megaproject, which generated great discontent among the general and professional public. The Academy of Serbian architects adopted a Declaration in 2015 with arguments against it. Belgrade Waterfront project was described as an attempt to look like Dubai by the Belgrade public. The civic Initiative Ne da(vi)mo Beograd (Do not drown Belgrade) was formed in 2015, and since then it has organized diverse activities and protests with the aim of keeping up the criticism of the Belgrade Waterfront project. In 2018, it developed into a political initiative and took part in Belgrade's city elections. Their activities started with Belgrade Waterfront project but are now embracing all issues regarding city policies and urban development. But the actual resistance of the public was very low.

It is unlikely that any costly investment in a megaproject in the city center would gain wide professional and public support, particularly in a country that is already over-indebted (Zeković et al. 2015). "In the post-socialist countries, transitional changes have created new power relations between different groups involved in urban development and increased influence of private investors" (Zekovic et al. 2018: 154).

The Agreement on Cooperation between the Republic of Serbia and the Government of the United Arab Emirates was adopted in 2013, after which the Serbian government founded the Belgrade Waterfront Company in 2014 with the aim of activating public funds for the construction of Belgrade Waterfront. The Belgrade Waterfront project has been the main topic in Belgrade since it was announced for the first time. In 2014, the project was integrated retroactively into the master plan of Belgrade (Amendments Master Plan 2014) while the Belgrade Waterfront Spatial Plan was adopted in 2015 (Lex specialis 2015).

Political and economic recovery started in 2000, but it is still far away from fulfillment. The main insufficiency is the lack of any clear transition strategy. No government since October 2000 has prepared a comprehensive strategy for transition. Since the democratic changes in 2000, Serbia has consistently demonstrated a desire to join the European Union. Serbia is officially in the process of joining the EU, after initiation of the Stabilisation and Association Agreement with the

European Union in 2007. In 2012, the country received candidate status. However, its integration into the EU depends on numerous factors (global trends, strategic thinking, and the recovery of its broken economy) and is still not certain.

In the last two decades, Serbia has lost almost its entire primary industry sector. The time of transition has not gone smoothly, and neither has the transformation of the planned economy into a free market economy. Accordingly, foreign direct investments have had a significant impact on the Serbian economy in recent years. There are many reasons for that, but mostly the aim is to increase access to international markets by improving the economic structure and obtaining new competitive qualities. The real and long-term challenges will always been hidden in economic and social change. Currently, Serbia considers China to be the most important financial and investment partner, as well as a leading partner in achieving its strategic economic goals. China is using current Serbian market openness. As a matter of fact, Serbia is at the center of *China's One Belt, One Road* infrastructure initiative, a transport and trade corridor running from China to Europe.

To quote Madžar (2017: 52), "*the realization of national goals, no matter how pledged and sublime they are, can be more or less successful only if the country equally successfully fits into its international environment and is well-oriented with respect to the objectives and the accompanying power of entities with whom it is in the inevitable interaction. To be successful, the small must adapt to the large.*"

There are still numerous problems. For example, there is a disproportion between the number of employees in the real sector and employees in the state sector. The bureaucratization of the state and agglomeration of regulations are still huge issues. Additionally, privatization, which is the base of any economy in transition, has acquired the status of being the biggest disaster in Serbia. It is not easy to transform from being a self-management system to a modern industrial society (Vujačić and Petrović Vujačić 2016). This process has been followed by corruption (Zurnić 2019), which remains one of the biggest problems in Serbian society. An excessive number of employees in state-owned companies, poor control of business and production processes by the party management, and awareness that the state will cover all losses are the characteristics of Serbian society at the beginning of the twenty-first century.

The Globalization and World Cities Research Network (GaWC 2018), an influential think tank on cities in globalization, included Belgrade on the list of global urban hierarchies for the first time in 2004. In that year, Belgrade was identified as a city with much evidence of the formation of a global city. Belgrade also maintained this position in 2008 (Taylor and Aranya 2008). Since then, Belgrade's ranking among global cities has progressed. In 2010, Belgrade positioned itself as a "gamma global city," while in 2012 it progressed to the position of being a beta global city. In 2018, Belgrade's ranking was again confirmed (GaWC 2018).

The architectural practice in this period is illustrated here through four examples: the Yugoslav Drama Theatre reconstruction at King Milan (Kralja Milana) Street, the National Bank of Serbia building at Slavija Square, the B2 business–residential building at Terazije Terrace, and Rajićeva Shopping Center in Knez Mihailova Street.

The building of the Yugoslav Drama Theatre (Figs. 1.36 and 1.37) was originally built for the needs of the Royal Equestrian Squadron in 1860 along today's King Milan Street (Figs. 1.38 and 1.39). Several decades later, the original intention changed when the National Theatre temporarily moved into the run-down building. Russian architect Nikolaj Krasnov was responsible for its reconstruction, after which the new theater started its operations in 1929, followed by a temporary change of use in 1931 when the building was adapted to suit the needs of the National Assembly. After the Second World War, in 1947, Serbian Momčilo Belobrk carried out the second adaptation of the theater. The building of the Yugoslav Drama Theatre became a modernist and monumental work, with a simple facade that has no ornamentation, and an entrance porch. After the Theater building burned down in 1997, a new reconstruction was necessary. The new reconstruction by Zoran Radojičić and Dejan Miljković completely changed the former facade, emphasizing glass as the dominant material. With the reconstruction of Krasnov's original facade and its incorporation into new the glass matrix, as well as retaining the basic ideas of Momčilo Belobrk in the entrance porch, the heritage is linked to a new modern architectural expression. The glass membrane over the new–old facade provides soundproofing from the noisy street. The renovated theater opened in 2003.

Fig. 1.36 Yugoslav Drama Theatre today, after its reconstruction in 2003. Photographer Biljana Arandelovic. Rights reserved

Fig. 1.37 Yugoslav Drama Theatre today, after its reconstruction in 2003. Photographer Biljana Arandelovic. Rights reserved

Fig. 1.38 King Milan Street. Photographer Biljana Arandelovic. Rights reserved

Fig. 1.39 King Milan Street. Photographer Biljana Arandelovic. Rights reserved

Fig. 1.40 The National Bank of Serbia building at Slavija Square. Photographer Biljana Arandelovic. Rights reserved

Fig. 1.41 The National Bank of Serbia building at Slavija Square. Photographer Biljana Arandelovic. Rights reserved

Fig. 1.42 Business–residential building B2, Terazije Terrace, Balkanska 2. Photographer Biljana Arandelovic. Rights reserved

The second example is the monumental building of the National Bank of Serbia (Figs. 1.40 and 1.41) at Slavija Square, built by Croatian architect Grujo Golijanin in the years of difficult political and economic conditions (1992–2005). The building has a distinctly postmodern design. The construction process was very drawn out for obvious reasons, following the events of Yugoslavian disintegration. The decision to construct it was made by the Council of Governors of the former Yugoslavia in 1976, while the detailed urban plan of the Slavija area from 1980 designated the exact location for the new building. The investment program was considered for 9 years and was finally adopted in 1998. With the collapse of the country, the National Bank of Serbia inherited all the problems and construction costs concerning the building. Construction began in 1992.

B2 business and residential building, built in 2010, is part of the Terazije Terrace area, a location of continual interest to architects and town planners, for which neither an architectural nor adequate urban design has so far been made. Ever since Belgrade became a modern city at the beginning of the twentieth century, this location has been the subject of studies, competitions, temporary designs, and various ideas. The best known proposal for this urban area is the award-winning design of famous architect Nikola Dobrović (1929–1930).

The B2 building (Figs. 1.42 and 1.43), designed by Slobodan Rajović, Zoran Nikezić, and Vasilije Milunović, is the result of a competition organized in the 1990s and is the first project implemented in this urban zone. The building is the first portal building of Terazije Terrace. B2 business–residential building can be

Fig. 1.43 Business–residential building B2 and Hotel Moskva. Photographer Biljana Arandelovic. Rights reserved

Fig. 1.44 Rajićeva Shopping Center. Photographer Biljana Arandelovic. Rights reserved

Fig. 1.45 View of Knez Mihailova Street from Rajićeva Shopping Center. Photographer Biljana Arandelovic. Rights reserved

Fig. 1.46 Rajićeva Street. Photographer Biljana Arandelovic. Rights reserved

Fig. 1.47 Knez Mihailova Street. Photographer Biljana Arandelovic. Rights reserved

Fig. 1.48 Knez Mihailova Street. Photographer Biljana Arandelovic. Rights reserved

considered to be successful, even though it is far from the monumental ideas of the Dobrović project. The project successfully worked out a cascading form that fits well into the urban area (Mitrović 2012).

The last example is Rajićeva Shopping Center (Fig. 1.40), a modern shopping mall of 15,000 m^2 opened in 2017 and located at Belgrade's most prominent site in the main pedestrian zone of Knez Mihailova Street (Figs. 1.44, 1.45, 1.46, 1.47, and 1.48). It was designed by Vladimir Lojanica with maximum consideration of the contextual conditions of the traditional city core. The completed project won a 1999 competition. Israeli company Ashtrom International is the project investor (the investment is worth 80 million euros).

References

Antić Č (2019) Srbija i sprski narod u 21. stoleću [Serbia and the Serbain nation in the 21st century]. Nedeljnik 21 februar 2019 54–55

Arandelovic B et al (2017) Belgrade: imaging the future and creating a European metropolis. Cities 63:1–19

Armenski T et al (2009) Perception of foreign tourists on the image of Serbia. Bull Serb Geogr Soc 89(1):39–63

Arsić M (2013) Da li je privatizacija glavni krivac za gubitak radnih mesta u Srbiji? [Is privatization the main culprit for the loss of jobs in Serbia?]. Kvartni Monit 34:71–73

Blagojević L (2003) Modernism in Serbia: the elusive margins of Belgrade architecture, 1919–1941. MIT Press, Cambridge

Bojović B (2003) Urbanizam Beograda u XIX i XX veku [Town planning in Belgrade in the XIX and XX century]. Godišnjak grada Beograda XLIX-L 109–130

Čomić Đ, Vićić S (2013) National and tourist identity of cities. The case study of Belgrade. Quasteus 2:15–27

Dimić Lj (1997) Kulturna politika Kraljevine Jugoslavije 1918–1941 [Cultural policy of the Kingdom of Yugoslavia]. Politika Beograd 24

Djurić-Zamolo D (1977) Beograd kao orijentalna Varos pod Turcima 1521–1867 [Belgrade as an oriental town under the Turks from 1521–1867]. Muzej Grada Beograda, Belgrade

GaWC (2018) Globalization and world cities (Research network). https://www.lboro.ac.uk/gawc/world2018t.html

Grčić M (1993) Razvoj industrije u suburbanoj zoni Beograda The development of industry in the suburban area of Belgrade. Zbornik radova PMF Beograd 41:259–268

Hadžić M (2002) Rethinking privatization in Serbia. East Eur Econ 40(6):6–23

Howard L, Allen D (2008) Visitor and non-visitor images of Central and Eastern Europe: a qualitative analysis. Int J Tour Res 10:27–40

Hirt S (2009) City profile: Belgrade, Serbia. Cities 26(5):293–303

https://zis.beograd.gov.rs/images/ZIS/Files/Publikacije/BUB_2018_E.pdf

Institute for Informatics and Statistics (2018) Belgrade in figures. Secretariat for Administration Sector for Statistics, City of Belgrade

Jakovina T (2011) The third side of the cold war. Faktura, Zagreb

Kulić V (2009) Land of in-between: modern architecture and the state in socialist Yugoslavia, 1945–65. PhD thesis, University of Texas at Austin

Kulić V (2014a) New Belgrade and socialist Yugoslavia's three globalisation. Int J Hist Cult Mod 2(2):125–153

Kulić V (2014b) East? West? Or both? Foreign perceptions of architecture in Socialist Yugoslavia. J Arch 14(1):129–147

Lex specialis—The Law Establishing the Public Interest and Special Procedures of Expropriation and the Issuance of Construction Permit for the Project Belgrade Waterfront. Official Gazette of RS, No. 34/2015

Lukić V, Anđelković-Stoilković M (2017) Interrelation of spatial disparities in development and migration patterns in transition economy: Serbia—case study. Hum Geogr J Stud Res Hum Geogr 11(1):65–76

Madžar Lj (2017) Geopolitical constraints on development policies of small countries. Acta Economica XV(26):9–54

Maksimović B (1983) Ideje i stvarnost urbanizma Beograda 1830–1941 [Belgrade urbanism ideas and reality]. Zavod za zaštitu spomenika kulture grada Beograda

Marković P (1992) Beograd i Evropa 1918–1941, Evropski uticaji na proces modernizacije Beograda [Belgrade and Europe 1918–1941, European influences on the process of modernization of Belgrade]. Savremena administracija

Marković JJ (2017) The image of Belgrade and Novi Sad as perceived by foreign tourists. J Geogr Inst Jovan Cvijić 66(1):91–104

Meriage LP (1978) The first Serbian uprising (1804–1813) and the nineteenth-century origins of the Eastern Question. Slavic Review 37(3):421–439

Mitrović M (2012) Arhitektura Beograda 1950–2012. Službeni Glasnik Beograd

Perović M, Žegarac Z (2000) The destruction of an architectural culture: the 1999 bombing of Belgrade. Cities 17(6):395–408

Petrović D (2001) Industrija i urbani razvoj Beograda [Industry and urban development of Belgrade]. Industrija 1–4:87–94

Petrović D (2006) Istorija industrije Beograda: razvoj i razmestaj industrije Beograda u XIX i XX veku [History of Belgrade industry: development and distribution of industry in Belgrade in the XIX and XX century]. Srpsko Geografsko drustvo Beograd

Petrović et al (2017) Urban geosites as an alternative geotourism destination—evidence from Belgrade. Open Geosci 9:442–456

Piljević Dj (2001) Beograd u XX veku: levica 1. Institut za noviju istoriju Srbije

Ratkaj I (2002) Saobraćajno-geografski položaj Srbije. Zbornik radova Geografskog fakulteta 50:33–46

Samardžić N (2015) Belgrade 1714–2014: Utopianism and Urbicide. In: Doytchinov G, Đukić A, Ioniță C (eds) Planning capital cities: Belgrade, Bucharest, Sofia. Verlag der Technischen Universität Graz, pp 212–224

Šuvaković M (2003) Impossible histories. In: Djurić D, Šuvakovic M (eds) Impossible histories: historical avant-gardes, neo-avant-gardes, and post-avant-gardes in Yugoslavia, 1918–1991. The MIT, Cambridge, Mass, pp 2–36

Taylor J, Aranya R (2008) A global 'urban roller coaster'? Connectivity changes in the world city network, 2000–04. Reg Stud 42(1):1–16

The Amendments on the Master Plan of Belgrade 2021 (2014) Official Gazette of the City of Belgrade, No. 70/2014

Topalović M (2011) Brick and gold: the urbanism and architecture of informal Belgrade, in Belgrade—formal/informal: a research on urban transformation. ETH Studio Basel Zürich, Verlag Scheidegger and Spiess

Vučetić-Mladenović R (2003) Evropa na Kalemegdanu: Cvijeta Zuzorić i kulturni zivot Beograda 1918–1941 [Europe on Kalemegdan. Cvijeta Zuzorić and cultural life in Belgrade 1918–1941]. Institut za noviju istoriju Beograd

Vujačić I, Petrović Vujačić J (2016) Privatization in Serbia—an assessment before the last round. Econ Ann LXI(209)

Willets P (1978) The non-aligned movement: the origins of a Third World alliance. Pinter Publishers

Weiss S (2007) Belgrade: Where subversion is normal. Perspecta 38; Yale Arch J (The MIT Press) (39):48–57

Weiss S (2013) National un-national. Natl Pap: J Natl Ethn 41(1):90–108

Zec M, Radonjić O (2012) Ekonomski model socijalističke Jugoslavije: saga o autodestrukciji [The economic model of socialist Yugoslavia: the saga of self-destruction]. Sociologija 54 (4):695–720

Zegarac Z (1999) Illegal construction in Belgrade and the prospects of urban development planning. Cities 16(5):365–370

Zeković S et al (2015) Spatial regularization, planning instruments and urban land market in a post-socialist society: the case of Belgrade. Habitat Int 48:65–78

Zekovic S, Maricic T and Vujošević M (2018) Megaprojects as an instrument of urban planning and development: example of Belgrade waterfront. In: Hostettler S et al (eds) Technologies for development, pp 153–164

Zurnić M (2019) Confronting corruption in Post-Milošević Serbia: discourse and institutions. Corruption and democratic transition in Eastern Europe. Political corruption and governance. Palgrave Macmillan, Cham, pp 75–106

Chapter 2
Urban Plans and the Transformation of Belgrade Through History

Milena Vukmirovic

Abstract This chapter provides an overview of Belgrade's most significant planning documents from the late nineteenth century to the present. These planning documents represent the core ideas and concepts that have guided Belgrade's urban development from 1867, the year marked as the turning point in the history of the modern urban development of Belgrade. It is the year in which the founder of modern Serbian urban planning, Emilijan Josimović, completed his urban plan for a modern European Belgrade and transformed it from an oriental town into a European city. Belgrade had only 25,000 inhabitants at the time of this reconstruction. Belgrade's history saw major changes in the nineteenth century, when everything related to the oriental Turkish Belgrade was demolished and the city was fully reconstructed. The planning documents are divided into four groups, from the end of the nineteenth century to WWI (4 documents), from WWI to WWII (3 documents), from WWII to 2000 (3 documents), and after 2000 (5 documents). The chapter explores various influences on the city's development and the strategic guidelines that can be seen in the chronological review of the planning documents presented.

Having an overview of urban plans is an important approach when monitoring urban development processes. Urban plans can be used to find trends and visions of development and to track various circumstances that have led to certain attitudes and ideas; they can also be of assistance with upcoming challenges. In the case of the city of Belgrade, we will try to observe the aspirations of its political elite to create a modern, European city through reading the basic conceptual urban concepts. The urban plans presented cover the period from 1867 to the present. This period has been associated with major crisis moments, such as the beginnings and endings of war conflicts and operations, changes in state regulation and legislative frameworks, and the expansion and reduction of state territory, which have all influenced Belgrade's development as the capital. To explain why the review of plans begins in 1867, two key events can be listed. The first event took place on April 19, 1867, when, after extensive negotiations with the Turkish authorities, they handed over the keys of the city to Knez Mihailo, after which the Turks officially

© Springer Nature Switzerland AG 2020
B. Arandelovic and M. Vukmirovic, *Belgrade*, The Urban Book Series,
https://doi.org/10.1007/978-3-030-35070-3_2

withdrew from Belgrade. In addition, this year launched the first legal regulation of the capital that occurred in parallel with Emilian Josimović making the first, comprehensive urban plan based on the model of European cities.

Taking into account different crisis circumstances, the overview of the plans for the city of Belgrade will be presented in four time periods (Fig. 2.1): (1) the period from 1867 to the beginning of the First World War (1914); (2) the period after the end of the First World War (1918) and the creation of the Kingdom of Serbs, Croats, and Slovenes until the start of the Second World War (1941); (3) the period after the Second World War (1945) and the creation of the Federal People's Republic of Yugoslavia (FPRY, later the Socialist Federal Republic of Yugoslavia—SFRY) until the wars in the former SFRY (1990); and (4) the period following the wars in the former SFRY, after 2000. Each of these periods is characterized by ups and downs, but such a general division is considered suitable to show the essential changes and key ideas defined by urban plans. These periods coincide with three different historical and developmental contexts (Vujošević and Nedović-Budić 2006; Zeković et al. 2016) characterized by the dominance of various political and socioeconomic systems. The first context includes a capital-based order in an underdeveloped agricultural state, and it lasts until the start of World War II. The second context coincides with the third period and is characterized by the development of an authentic socialist system and a legislative structure of the state that ends with the breakup of Yugoslavia and the collapse of the socialist system. The third context involves the stage of the disintegration of Yugoslavia and the collapse of the socialist system (1990–2000), in parallel with the post-socialist transition of society and the economy within a capitalist system of neoliberal discourse. It coincides with the fourth urban planning period, from 2000 to the present.

2.1 The First European Patterns of Belgrade's Urban Development from 1867 to 1914

Four urban plans were developed in the period from 1867 to the beginning of WWI. The first urban plan was entitled *Explanation of the proposal for urban regularisation of the part of Belgrade which lies within the Moat* by Emilijan Josimović in 1867. With this plan, organized urban planning practice began in Belgrade. The second plan, from 1878, was developed by Stevan Zarić for use by Belgrade's primary schools and was considered to be the most important, most complete, and accurate plan of Belgrade in the nineteenth century. Bešlić's plan (the third) was designed in 1893 following the *Decision on the Regulation of Belgrade*, and it represented an expansion of Belgrade's urban area. The fourth plan from this period, adopted in 1908, was made by French architect Alban Chambon, and it was also the first Master Plan of Belgrade.

On the territory of Serbia, regulatory plans were made on the basis of the Law on Towns (1866, amendments from 1885), which obliged all cities and towns to

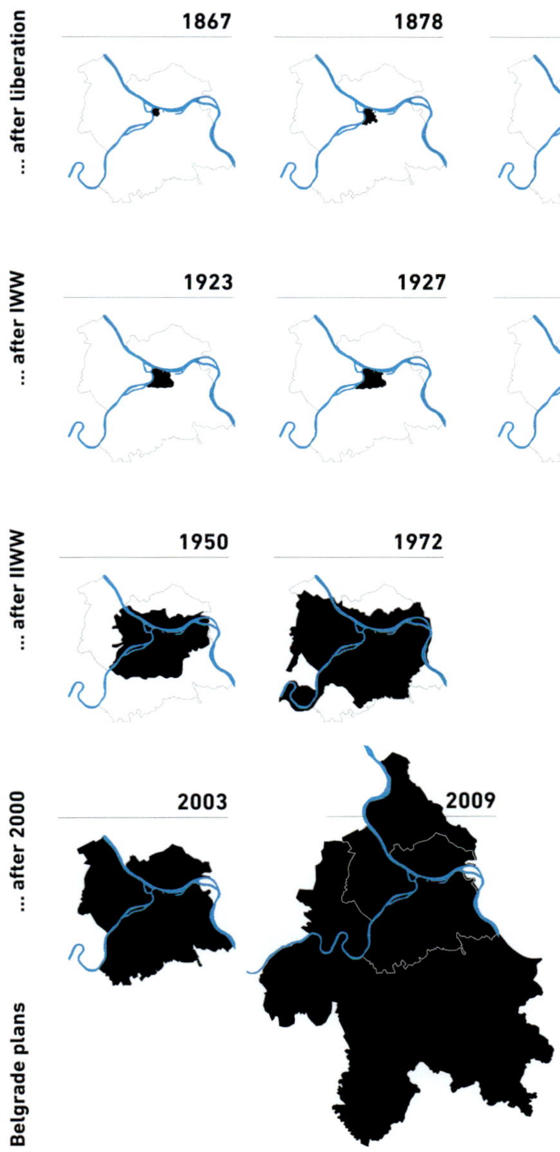

Fig. 2.1 The territory covered by Belgrade urban plans from its liberation 1867 to the present

develop a proper regulation and leveling plan within 3 years of its adoption. Plans were made for many years following this, because a length of 3 years was very ambitious, given the few technical personnel of that time (Vusanović Macura 2014, p. 34). According to Vuksanovic-Macura (2014), these plans were very simple, and often, in planning terms, they had a control line, a construction line, a building's position in relation to adjacent objects, and a plot size. Therefore, the need for a unique, state-building law that would deal with the matter of planning settlements and their construction was present everywhere. The adoption of the construction law preceded the adoption of Yugoslav Technical Terminology (1920), the publication of the Construction Law Project (1925), and the Rulebook for the preparation of regulation and leveling plans (1925).

The Construction Law for the City of Belgrade was made in 1896 and was amended in 1898 and 1901. This law relied on the Construction Rulebook for the City of Belgrade, made in 1897 (Vusanović Macura 2014, p. 36). These acts were aimed at intensifying construction activity through certain tax exemptions for new construction and the upgrading of floors, while some of the regulations concerned the formation of a regulation lines and leveling. The city was divided into building zones made up of dense settlements, central settlements, and sparse settlements (Obradović 1932), which was the principle used until the end of WWII and the adoption of new laws.

The year 1867 is marked as the turning point in the history of Belgrade's modern urban development. It is the year when Emilijan Josimović, the founder of modern Serbian urban planning and the first Serbian urbanist, completed his inventive urban plan of modern European Belgrade, which was actually a study that he had worked on as an individual enthusiastic researcher. Today, we refer to his work as the first modern urban plan of Belgrade, but in reality, it was actually Josimović's book on the reconstruction of Belgrade within the Moat that made the biggest impact in which he explained his proposal for new urban regulations with a plan to the scale of 1:3000 (Fig. 2.1) after having recorded the Turkish, oriental Belgrade. Josimović's book was a kind of urban revolution (Terzić 1997; Medaković 1997).

According to Blagojevic, "Josimović was a contemporary to the famous European urbanists Haussmann, and Cerdà, and his plan followed soon after their famous plans for the reconstruction of Paris (1853–71), and Barcelona (1859–70). His plan demonstrates the universality of the European planning paradigm of the period, and more precisely of the ring principle, which was employed in Vienna, yet it clearly brings out the specific character of Belgrade into its proposals for reconstruction" (Blagojević 2009).

Josimović's work is of historical importance. The proposals and ideas included, for example, connections between the fortress and the city through the construction of a tunnel under the Belgrade ridge, a ring of parks around the Moat and others. Josimović proposed the transformation of the Kalemegdan area into a public park. In addition to the old, narrow roads entering through three gates, Josimović created three other passages into the new town.

This new link between Knez Mihailova Street and Terazije Square emphasized the importance of Knez Mihailova and transformed it into a new trade center. The tracing of Knez Mihailova Street played a crucial role in the creation of a new commercial and social center in the old town. As it intersected with the old bazaar (King Peter Street (Kralja Petra)), this new, wide street was a modern extension of the old bazaar. At the other end, this street led to Terazije Square, which was a new commercial, social, and administrative center.

Being aware of the importance of the Danube, Josimović proposed the construction of an organized quay around the city, along the Danube. His most daring proposal was to dig a tunnel under Kalemegdan, which would provide a short and convenient connection between the Sava pier and the Danube wharf. Another proposal was to surround the old town with a circle of parks, and to connect it with newly planted tree-lined streets, from the Sava to the Danube, with a total surface area of 50,000 m^2 (Terzić 1997; Medaković 1997; Maksimović 1967). Blagojević (2009) compared Josimović's plan to Vienna, noticing the "clear influence of forming a ring zone of green and open urban character in place of the demolished ramparts. Josimović, however, departs from this well-known model and introduces what he calls the *town wreaths/ghirlanda della città/gradski venci*." She further points out that "Josimović clearly sees Belgrade's reconstruction not as copying the imperial European capitals, but as creating an idiosyncratic urban structure incorporating parks and green town wreaths, which will be tailored so as to provide beauty, satisfaction and a healthy environment to its citizens."

According to Josimović's plan, only around a hundred buildings would be retained. Josimović reserved a dominant position in the plan for the National Theatre edifice, on the new square around the ex Stambol gate, in the space which is today occupied by the National Museum building. Reconstruction of the town within the Ditch during the 1870s and 1880s, after the demolition of the old buildings and tracing of new, wide streets, encouraged the construction of housing and commercial buildings due to the favorable price of the building land, and due to the proximity of the old city center.

His ideas were a fundamental turning point both in theoretical and practical terms, as he established a basic plan for building the modern transformed Belgrade, which was actually implemented (except for some details), but at the same time he determined a standard for the reconstruction of other towns in Serbia. The center of Belgrade was developed on the basis of Josimović's plan until the end of the nineteenth century. However, there was also a review in that period of some of the concepts in the plan and those changes were made in the two plans that followed: Zarić's plan from 1878 and Bešlić's plan from 1893.

Zarić's plan from 1878 (Fig. 2.2a) was the first regulatory plan of Belgrade for the construction area of that time, which lay to the south of Slavija Square, its most extreme limit being Belgrade (Beogradska) Street, including some parts of Palilula up to the present-day Dalmatinska Street. It also included the present-day Savski Square which is, in the plan, marked as the Venetian pond (Bara Venecija). This plan clearly shows that some streets were intersected and regulated according to Josimović's plan, while the Moat was overlaid. Zarić's plan shows Belgrade as a whole, almost four times bigger than the oriental Belgrade in the Moat.

Zarić's plan completely defined the orthogonal structure of Vračar, which is today the most densely populated central municipality. The upper part of the Danube slope was regulated by the plan, while the lower part, close to Cara Dušana Street, still had the old system of irregular streets. The former Palilula village was

Fig. 2.2 Plan of Regularization of Belgrade, scale 1:3000, 1867. Author Josimović (1997)

now an integral part of the city's territory, with its sharp-angled street network and large, irregular blocks. Kalemegdan was represented by dense networks of twisting paths in the style of an English park. The plan included a significantly wider territory than Josimovic's plan, with new important facilities and arranged spaces: the railway station, two planted and arranged town parks—Kalemegdan and the park at Knez Miloš's former residence, on the corners of Knez Miloš and Nemanjina streets. The old town within the Ditch was at that time almost

(a) (b)

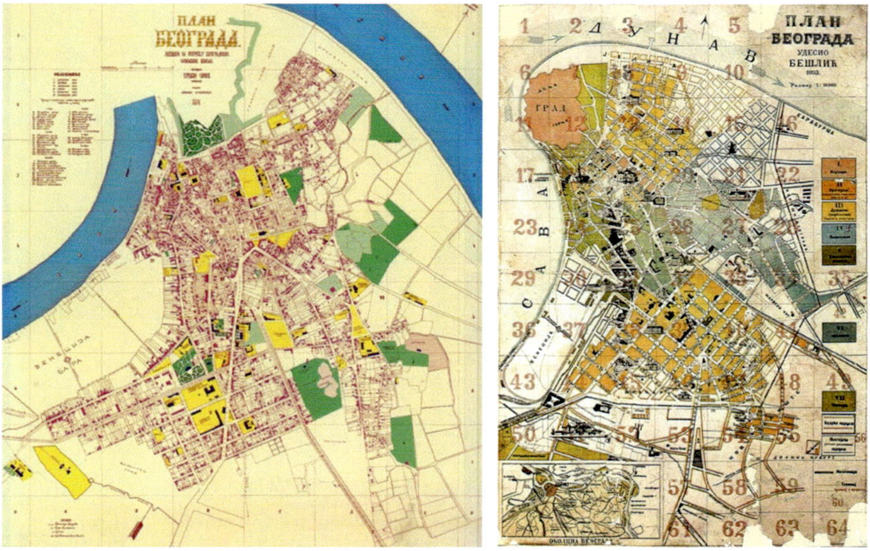

Fig. 2.3 **a** Plan of Belgrade made to meet the requirements of Belgrade's primary schools, 1878 and **b** Plan of Belgrade drawn up by Jovan Beslić, 1893. Archive: Belgrade Urban Planning Institute

completely reconstructed, with a rectangular street network and relatively small blocks (Maksimović 1983).

Jovan Bešlić's plan from 1893 (Fig. 2.3b) came after the adoption of the *Decision on the regulation of Belgrade* from 1891, which was at the same time also the first legal decision for 223 open public urban spaces (streets and squares). The document shows the new expansion of the city area with a clear presentation of the surroundings of Belgrade at the time. In order to present individual zones of the city, the plan was designed in a number of colors (highlighting Belgrade Fortress, Kalemegdan Park and the districts: Dorćol, Varoš, Savamala, Palilula, Terazije, Vračar, and Topčider). The author also used the strip map technique for showing the most significant buildings in Belgrade. Railway and tram lines were shown on the map. Because of its decorativeness and clarity, this plan is considered to be the first tourist plan of the city. The green belt and alleys, the distinctive elements of Josimović's plan of Belgrade in the Moat that divided the inner from the outer city, completely disappeared. The 1893 plan of Belgrade represents the first regulatory plan of the whole city. Bešlić's plan has a special importance, because it expanded Josimović's plan of the Danube slope to a much larger space from Dorćol and Palilula to the Danube banks.

French architect Alban Chambon made the first general plan of Belgrade in 1908. The plan was clearly framed by an arterial ring road. Chambon's plan was significantly different from the previous ones, because it anticipated the

(a)

(b)

Fig. 2.4 **a** National assembly and complex of Ministry buildings and **b** Proposal for St. Sava's Temple. Archive: Vukmirović

arrangement of a range of social buildings and several architectonic compositions. The basic idea of the city's reconstruction followed the principles of French academic urbanism. According to Arandelovic et al. (2017), "a circular boulevard clearly separated the construction area from the rest of the city, while the formation of eleven ensembles of buildings aimed to strengthen the identity of Belgrade and bring it closer to the appearance of other European capitals." The 11 ensembles of buildings in the plan were the Temple of Saint Sava in Savinac, the Triumphal Arch in Slavija Square, a complex of royal Serbian museums, a complex of ministry buildings and the National Assembly, the Royal Court Complex, the development of Terazije and its terraces, a post office, a medical faculty, a stock exchange, municipal buildings, and a courthouse. Although not built as a result of this plan, the idea was for them to form magnificent open public spaces (Fig. 2.4).

Chambon was the creator of a number of brilliant public space proposals and perfectly emphasized stretches of Kalemegdan, Slavija, and King Alexander Boulevard (Roter-Blagojević and Radivojević 2007). His vision of Terazije Terrace is significant, connecting Terazije Square with Queen Natalija Street, forming a number of smaller squares, where the principle of *hanging gardens (jardin suspendue)* is applied. He will be remembered as the first to create this central city stretch of Terazija plateau toward the Sava waterfront. This was his largest contribution, but unfortunately it was not implemented due to the outbreak of war (Balkans wars 1912–1914 and then WWI). All of these interventions were envisaged in the central city core, when at that time 83% of the buildings were single storey and not many of them were built with solid materials. Although revolutionary for the time and in line with the modern trends in other European capitals, the resistance of the Serbian public toward this plan was extremely strong. Chambon was considered harmful and a threat to the local culture. Because of this attitude, which the prime minister and king went along with in the end, the plan had no chance of realization.

His proposals are still relevant today and the most famous example of this is Terazije Terrace. It is interesting to mention that Chambon's plan was resuscitated in 2003 on the initiative of Professor Miloš Perović, when students from the Belgrade architectural school reinterpreted some of Chambon's basic ideas in a contemporary spirit in their final projects. This project was presented at the Venice Biennale of 2006. In this way, it was shown that some elements of the Chambon concept can be used as the basis for modern solutions, especially at sites that have not received their definitive form over time.

It is important to note that this period was characterized by constant political quakes—shifts in the government, elections, conspiracies, abolition of the constitution and adoption of a new one, the introduction of a personal royal regime, etc. This turbulent period ended with the upheaval that took place on May 29, 1903, with the murder of the Obrenović royal couple and the arrival of the Karadjordjević dynasty. However, although according to Ristović et al. all these events kept the country paralyzed, it was in those years that Serbia began to change and evolve, especially its capital. In Belgrade, huge infrastructure works were launched (water supply, sewage, electrification, tramway traffic, street decoration, etc.) in order to bring it closer to the European cities that were its model. The encouragement to start these interventions, on the initiative of the president of the Belgrade municipality Dr. Vladan Djordjević, took place on October 17, 1884, when a decision was made at the meeting of the committee of the Belgrade municipality to begin the most extensive works in the history of the city. A three-member delegation from Belgrade was sent to visit the most important European cities in order to study the functioning of the communal, traffic, and urban solutions there, as well as to see the infrastructure. Their main destination was London, but they also visited Folkstone, Hastings, Eastbourne, and Brighton. On their way to Britain, the delegation visited Timisoara, Budapest, Vienna, Munich, and Paris. After returning from the trip (January 26, 1885), an exhaustive report was made containing detailed proposals of the measures that the municipality should carry out in order to make Belgrade a modern European capital in the communal sphere (Rastović 2013).

Although many large works were launched, work on them was often suspended, mostly due to political problems. "There was disagreement in the municipality itself, but also between the administration of the city and the government, because Belgrade was mostly under the authority of the opposition, which was a disgrace to the state authorities". However, Herbert Vivien, summarizing the impressions of his visit to Belgrade in the nineteenth century, wrote that it "is amazing what prosperity the city has experienced since the Turkish times" (Vivien in Rastović 2013).

After the change of monarchy, Serbia entered a more peaceful political period; however, internal problems were replaced by foreign policy problems (sanctions, a customs war, annex crisis, and others). Although foreign policy conditions were difficult, Belgrade continued its path of urbanization and Europeanization (Ristović et al. 2008). Most of the planned works were completed in 1911, such as the construction of Kalemegdan, central streets and Terazije, an extended water supply system, a sewage system, an extended tram network, and so on. However, soon after that came a period of "hibernation" that interrupted the developmental flow, when the capital was subject to bombing and ruin.

2.2 Plans of Belgrade in the Period After WWI and Before WWII

After the First World War, a legislative framework was developed in order to define the planning regulations that would outline the development path of the capital of the newly established state—the Kingdom of Serbs, Croats, and Slovenes. Among the first initiatives was the adoption of the Financial Law in 1922, which included the directive on the regulation of Belgrade, on the basis of which the Municipal Committee drafted the Technical Requirements. In this way, the amendments to the Construction Law for the City of Belgrade were consolidated, aimed at encouraging the construction of multi-storey buildings. In the same year, the Commission for drafting the General Plan of Belgrade and the Commission for drafting the New Construction Law for the City of Belgrade were formed (Obradović 1938). An incentive for arranging the whole territory of Belgrade was also the Law on territory adopted in 1929. The Law aimed to divide the city territory into construction zones and to specify the method of construction in each of the zones, thereby putting the spreading of Belgrade's periphery under the control of construction policy and regulating further development.

However, the key stimulus for the regulated and planned arrangement of Belgrade was the Construction Law from 1931. The aim of this law was to open the way for cities and municipalities to take the necessary measures to develop and improve their municipality, whether urban, rural, or any other type of settlement (Obradović 1938). The Law was supplemented by the Instructions to Municipal Authorities for the Application of the Construction Law, which contained the duties and rules of the municipal administrations. The main job that the municipality was supposed to do was to adopt the arrangement bases, consisting of the regulatory plan, the regulations for its execution, and the building guidelines. However, it was difficult to apply the Law due to a lack of funds for drafting regulations and a lack of personnel, equipment, and technical services.

A competition for the General Plan of Belgrade 1923 was announced in August 1921 in line with the modern European tendency in city planning. The competition program focused on the rapid development of the surrounding settlements and emphasized it as a major problem which required an immediate solution. The aim was to form a new General Plan that included new settlements that would be attached to the city. The participants in the competition were obliged to consider and include the areas of Avala and Rakovica. After the competition ended, none of the 22 proposals received had produced a complete solution to all the problems given. However, after seeing some ideas for Belgrade's territorial expansion toward the west and toward Zemun (which would be possible after the construction of bridges and docks), the conclusion of the jury was that Belgrade and Zemun, until then two separate cities, could merge into one. Belgrade's expansion on the left bank of the Sava River was mentioned here for the first time (Vuksanović-Macura 2014). The low cost of land on the periphery of the city resulted in the trend of building family houses there (Voždovac, Dušanovac, Pašino Brdo, Banovo Brdo, and Dedinje). However, as these parts of the city were not included in the general plan, the construction proceeded on arbitrarily divided lots, and streets were traced without plans.

The committee stressed that the most important task of the general plan was the division of the city into zones, with a predetermined building system and population density, as opposed to the inherited chaos. Among the major advantages of the 1923 General Plan was the creation of new social centers surrounded with green zones. It is also this plan which proposed the introduction of apartment blocks along the right bank of the Sava near the train station and tracks. Some contemporaries consider this as the beginning of the city's formation within the Sava amphitheater. Requirements of the General Plan 1923 emphasized the creation of regular architectonic units, squares, with very prominent assemblages around the Parliament buildings and in the area of St. Sava Temple (Maksimović 1983).

According to the agreements from the Peace Conference held in Paris in 1919–1920, Belgrade, as the capital, needed to expand and connect with Zemun, which was the idea in accordance with international plans on the formation of the so-called Greater Belgrade. During this period, the most significant change was the elimination of the former border on the Sava, which made construction on the left bank of the river possible. That direction of expansion was natural for Belgrade and was in accordance with the new needs of Belgrade as the seat of the new territorially enlarged state. The 1923 plan included the treatment of New Belgrade as part of the urban fabric. Two competition entries attracted attention during this period: one was *Singidunum Novissima,* which focused on the park and fair complex in New Belgrade, while the other one *Urbs magna* proposed installations on the quay in New Belgrade. These were the first drawings of New Belgrade.

The new plan of 1923, made by Djordje Kovaljevski, included Zemun, which was integrated into the potential tissue of New Belgrade and War Island (large recreational sports area of Belgrade). Two bridges were also envisaged by the plan, one toward War Island and the other a railway bridge over Ada Ciganlija (as well as many boulevards modeled on Paris and Brussels). The plan also anticipated installations on the quay and a large international exhibition space. This new concept of development on the left bank of the Sava can be interpreted as "the ambitions of the ruling elite that at least on paper show the progress of the young state and evidence its belonging to developed Europe" (Djordjević and Dabović 2010). However, Kovaljevski's approach could be characterized as "traditionalist" because it was modeled on the imperial plans of major European cities, at a time when in European centers individual urbanists were asking questions about how to justify city planning according to artistic principles and were developing alternative models. For various reasons, about which we can now only speculate, all of these

(a) (b)

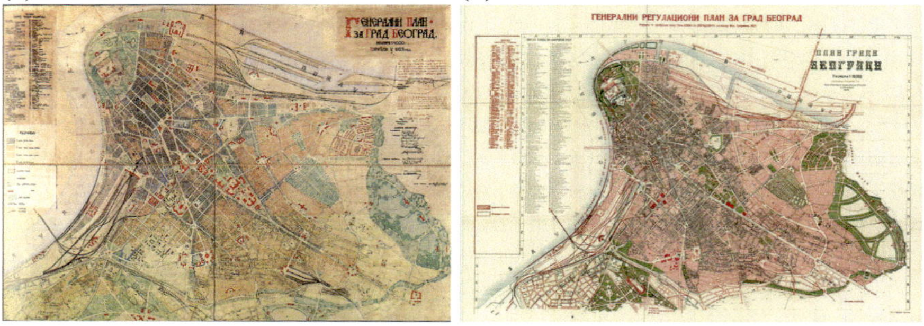

Fig. 2.5 **a** General Plan of Belgrade 1923 and **b** General Regulatory Plan 1927. Archive: Belgrade Urban Planning Institute

ideas were abandoned in the same year that they were presented, but their revitalization became popular after WWII.

The General Regulatory plan of Belgrade (Fig. 2.5b) was made by engineer Jovan Obradović in 1927 as a comprehensive proposal and a continuation of the tendencies and ideas initiated by the plan from 1923. It is important to emphasize that the plan from 1927 was the first to give a complete list of streets in alphabetical order. In addition to public buildings, monumental stretches, boulevards and functional roads, ports, and railway installations, the plan also envisaged the construction of garden colonies and established the new position of the Pančevo Bridge.

Garden colonies were the idea of modernist architect Jan Dubovy and were proposed as an alternative to the plan of 1923, given the projection of "the future of Greater Belgrade" with an area of about 100 km^2, developed in a radius around the historical center on the ridge in two directions. The first direction was the lowering of Belgrade on its rivers, while the second was the integration of the city and surrounding villages (Žarkovo, Rakovica, Banjica, Jajinci, Kumodraž, Veliki and Mali Mokri Lug, Mirijevo, and Višnica). Dubovy treated the area of Bežanija, which is later where New Belgrade was developed, as part of a broader periphery, a garden city immersed in greenery. However, the concept of a garden city remained mostly unrealized, only partially accepted in the planning of residential areas of more middle-class family homes and villas. These settlements were built in new locations outside the inner city center and in the areas of Dedinje, Senjak, Topčider Hill, and Kotež Neimar (Djordjević and Dabović 2010). The General Regulatory plan from 1927 gave an unrealistic program of reconstruction and expansion of the city and most importantly, it was not able to solve the growing problem of poverty among the population that had emerged due to the speculative housing market that characterized the period.

The exit of Belgrade onto the river and orientation for the further development of the city onto the left bank of the Sava can be further seen in the first prize winning competition design of Nikola Dobrović, in his solution for the Terazije plateau from

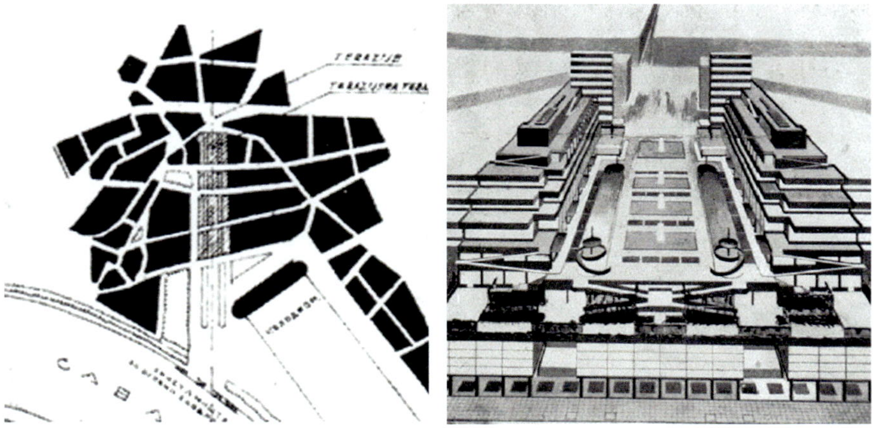

Fig. 2.6 Terazije Terrace—the first prize solution for Terazije Plateau from 1929 (Perović and Krunić 1998)

1929 (Fig. 2.6). Dobrović wanted to highlight the natural connection between Terazije and the banks of the Sava by forming terraces with gardens, pools, and fountains along the route. Dobrović's competition project continues to be taken as the basis for the future transformation of this area because it emphasizes one of the key city directions of movement through the city, making a natural gate to the river. To this day, except for one segment (B2 business–residential building featured in the first chapter), the idea has not been implemented. The significance of this solution is in the break with traditional planning and the creation of the image of a modern city that strives toward rapid transformation. This idea was comparable with entries for other competitions of the time for Alexanderplatz in Berlin and the Rockefeller Center in New York in 1928 (Perović 2003).

The General Regulation Plan of 1939 served as the basis for issuing a new international competition for the Master Plan of Belgrade, which was not implemented in the end. The main role of this plan was to address illegal, unhygienic settlements that developed as a ring around the border of the previous plan. As a consequence of the growth in its illegal settlements, Belgrade, as the administrative and political center of the new state, rapidly began to see an increase in its number of inhabitants.

In May 1933, according to the new regulations, Belgrade was divided into 20 construction zones, and in August, into 18 stretches. However, the most significant expansion of Belgrade occurred in March 1934, as a result of political motives to "make Belgrade the priority as the capital of a large European country" (Gašić 2011). Belgrade's new municipalities were determined, after which Zemun and 13 new municipalities on the other bank of the Sava and Danube entered into greater Belgrade (Zemun, Višnjica, Mirijevo, Mali Mokri Lug, Veliki Mokri Lug, Kumodraž, Jajinci, Resnik, Kneževac, Žarkovo, Železnik, Bežanija, Ovča, and Borča).

Fig. 2.7 King Alexander I Bridge. Archive: Fotografije starog Beograda

Although the first construction that connected Belgrade's two shores was a railway bridge, built upon the decision of the Berlin Congress in 1884, the construction of King Alexander I Bridge (Fig. 2.7) and its opening to traffic on December 17, 1934, was the first event on the left bank of the Sava. Both of these ideas, the bridge being realized in full, had as their goal the affirmation of the city territory on the other bank. The bridge was opened by Prince Pavle immediately after the assassination of King Alexander I Karadjordjević in Marseilles on October 9, 1934. As part of the city's defensive tactics in World War II, the bridge was demolished by local engineers on April 9, 1941 (Ilijevski 2013).

King Alexander Bridge opened up the possibility of developing the Old Fairground, marking the beginning of construction on the left bank of the Sava (Fig. 2.8). In addition to the economic importance of the fairground, it was also a modern spatial composition which stood out in its appearance in comparison to other parts of Belgrade. The framework for the implementation of the Old Fairground project was defined in 1936, the same year that Belgrade was promoted as a tourist destination in the Kingdom of Yugoslavia. The basic idea of the Old Fairground project included the radial positioning of pavilions around a central tower. The tower had two basic functions: it was both an exhibition pavilion and an urban landmark that could be seen from different parts of the city. The first

Fig. 2.8 Belgrade's Old Fairground. Archive: forensic-architecture.org and Vukotić-Lazar and Đokić (2006)

exhibition was organized in the autumn of 1937, with 7 major international and more than 20 private pavilions (private firms and individuals). Over time, the purpose of this space changed, and after the occupation in 1941, it became a concentration camp. After the war, the new government, both communist and anti-fascist, was doubly frustrated regarding this space. According to Vukotić-Lazar and Djokić (2006), "this location had a tragic (during the War) and capitalist (before the War) past, so the complex was absolutely ignored in the period of the postwar renewal." The function of this space has still not been defined. There are many plans but nothing final.

This period was also characterized by political affairs, reversals, changes, abolition of the old and establishment of the new constitution, and the beginning of an authoritarian regime, which culminated in the killing of King Alexander Karadjordjević in Marseilles, on October 9, 1934. At the same time as these internal problems, there were crisis events on the world scene (Russian revolution, world economic crisis, and the strengthening of totalitarian regimes and fascist orders), which also reflected on the situation in the country and its capital. In these difficult times, Belgrade managed to concentrate on its own development and to record one of the biggest transformations in its history. Broad boulevards and representative buildings were built, while the "life of the inhabitants was more like life in any European capital of the time" (Ristović et al. 2008, 010).

It is interesting to mention the opinions of British travel writers (David John Footman, Lena A. Yovichic and Lovett Fielding Edwards), who analyzed the changes that took place in Belgrade during the 1930s. The authors listed the following characteristics of this period: the speed at which the old buildings were replaced by new ones, unsustainable construction, unsystematicity, pretentiousness of the architecture, pretentious modernity in contrast to the unfinished process of asphalting and paving of streets, rapid expansion beyond the original boundaries, etc. (Lazarević-Radak 2011, 240). In relation to the general character of Belgrade, so-called hybrid forms prevailed in its structure that combined traditional and modern elements, while despite the existence of plans and laws, the city developed partially and unevenly. Based on the above, we can conclude that the plans and vision for the development of Belgrade that were created during this period were too ambitious and fairly ahead of their time, as well as beyond the existing capabilities. The beginning of the Second World War led Belgrade into a new phase of "hibernation," when it was bombarded and demolished several times during the war.

2.2.1 Planning After WWII up to the Beginning of the Yugoslav War

After the Second World War, the state order changed, the monarchy was abolished, the republic was introduced, and a new administrative-socialist system was established in FNRJ, so the period up to 1950 was characterized by the adoption of several new acts that defined the regulations in the field of managing construction land (Zeković and Vujošević 2018). In this way, the legal continuity of the Kingdom of Yugoslavia was abolished. Significant Acts adopted were the Decree on the Registration of Ownership Rights on State Real Property (1947) and the Law on Confiscation, Nationalization, and Expropriation, which was the basis for the transfer of property from private and other forms of ownership to state ownership. In this way, the dominance of state ownership was established. In the post-war years, reconstruction and the preparations for the new General Plan of Belgrade began, preceded by the introduction of the Essential Provisions for Design and Construction (1948) and the Basic Regulation on the General Urban Plan (1949).

According to Zeković and Vujošević, the period from the 1960s onward was characterized by an authentic socialist governing system that was initiated by the laws of 1950–1951. The key focus of this system was socially owned property as a unique form in the world. The 1963 Constitution introduced self-management in all sociopolitical communities, while the reforms from 1964 to 1967 and 1969 reduced the role of the state in the economy, facilitating foreign investment and the conditions for the development of market socialism. “The model of self-management has promoted decentralization and strengthened the bottom-up approach to decision-making, planning and management” (Zeković and Vujošević 2018, p. 42). This period saw the complete transfer of construction land to the state, thereby giving the state full control. In 1961, the first Law on Urban and Regional Planning was adopted in Serbia, which established the basic legal terminology and legal nature of the plans. The 1974 Constitution introduced social planning for sociopolitical communities that were obliged to determine policies, guidelines, and measures for the implementation of these plans. According to Zeković and Vujošević (2018), at the end of the 1980s, social planning disappeared, and the arrangement of space was directed to municipalities. At the same time, a more powerful socioeconomic system of planning and practice, in accordance with the concept of associated labor and the agreed economy, contributed to urban and spatial planning gradually losing significance.

According to Vujošević, the observed period of almost five decades can be divided into three time frames: (a) central-command planning, from 1947 to 1965; (b) political decentralization and social self-management planning, from 1965 to 1989; and (c) “democratized” planning, from 1989 to the present, bearing in mind the social and political context in which the plans were created and realized.

After WWII the sociopolitical conditions changed, which caused a change in the status of Belgrade as the capital of the new state. For these reasons, a search began for a place where Belgrade would continue to build and develop. The conceptual changes can be seen in the selection of the location for the construction of the new city—Bežanija field, which was a place “free of history.” This was a completely clean area on which no urban structure had ever been formed. The idea of building a new part of the city on the left bank of the Sava was also the main feature of the first post-war Master Plan of Belgrade, which was adopted on October 20, 1950.

Fig. 2.9 Empty territory of New Belgrade on map from 1961. Archive: Military Geographical Institute (left) and the Masterplan of Belgrade 1950 (right). Archive: Belgrade Urban Planning Institute

The plan had as its starting point the Preliminary Master plan of Belgrade from 1948 by Nikola Dobrović, the recommendations of the Five-Year Development Plan of Belgrade (1947–1951), and the Program for the future development of Belgrade from 1966. The important basic assumptions of the plan were growth in the number of inhabitants to one million, the equality of living conditions over the whole territory of Belgrade, and construction of a new part of the city on the left bank of the Sava, with the aim of Belgrade becoming a suitable political, economic, and administrative center. These goals were the basis for future significant changes in the nature and urban character of Belgrade, which were possible for that period using the progressive urban planning principles defined in the *Athens Charter*. The city model for the urban development of Belgrade was developed in line with the theory of Le Corbusier (emphasizing monumentality) (Fig. 2.9).

The construction of New Belgrade after WWII was a major urban project in the former Yugoslavia, which resulted in an influx of new residents from within Serbia. The first phase of New Belgrade's development was the extensive construction of housing, which began in the post-war period (more in the New Belgrade chapter). A new style of housing policy was put into practice in the building of New Belgrade. The collective ownership concept in socialism had the ideological assumption that a flat was an asset of the society (as was, for example, greenery or the city infrastructure). The New Belgrade zone was divided by the main traffic routes into four longitudinal belts, the one closest to the riverbank was intended for the administrative and governmental institutions, and the remaining three for housing. A zone for special construction was planned in a transverse direction between the buildings of the Federal Executive Council and the future railway station in New Belgrade. The Great War Island was connected with the mainland, and a channel of the Danube was converted to a lake (Stojanović and Martinović 1978; Petričić 1976).

The New Belgrade concept and plan radically changed over the years. The construction of New Belgrade was in full swing the middle of the 1960s when the

Fig. 2.10 General urban plan of Belgrade 1972. Archive: Belgrade Urban Planning Institute

new city gained the principal outline of its present-day appearance. With this change, a large number of initially planned ministry buildings and governmental and party edifices were excluded, and housing construction on a massive scale underlined this new socialist city development.

The stabilization and development that the country achieved during the 1970s and '80s could be noticed in Belgrade as the capital. During the '70s, due to the large population inflow, Belgrade reached over one million inhabitants. Faced with new development challenges (with goals that proved to be unrealistic), the need for a new General Urban Plan appeared. The decision to make a new plan was made in the second half of 1969, and the plan was adopted on March 23, 1972 (Fig. 2.10).

General urban plan of Belgrade for 2000 (1972) was produced to cover a period of 30 years, with the aim of transforming Belgrade into a modern European capital. The territory of the city was divided into the center, wider center, and suburban areas. The plan envisaged changes in the parameters that defined the population density and the ratio of free and built areas. During the 70s, a more moderate influx of people could be seen, as well as intensive collective construction. But despite the existence of regulations, illegal settlements were still being built, with Kaludjerica as the most famous such example. One of the largest enterprises in this plan was to build the recreational area of Ada Ciganlija.

The plan was formed on the conclusions of more than 150 documents and studies. Forecasts showed that in the year 2000 Belgrade would have about 2,000,000 inhabitants who would live in agglomerations of settlements over the whole administrative area of the city, and which would cover around 70,000 ha.

The area included the territory from Batajnica in the west to Boleč in the east and from Krnjača and Pančevo in the north to Avala, Sremčica, and Umka in the south (GUP 1972). In that area, the planned changes were as follows: the existing villages and settlements around the city would become part of a single city agglomeration and lose their rural and *suburban* character; the city would spread to the surrounding terrain, not continuously, but in the form of *archipelago settlements in a sea of green* (by which Jan Dubovy's ideas would be reanimated); and the expansion of the city would unfold following the directions of linear expansion to Smederevo, Avala, Obrenovac, and Batajnica. Expansion in this way was supposed to provide a linear organization within new residential settlements, and centers could rely on a system of longitudinal roads or rails leading to the central zone.

Although the plan envisaged the construction of new settlements, the central zone was expected to remain a symbol and place that would be easy to visit and enjoy "not only for the purpose of work or cultural and artistic events, but also for meetings with friends and enjoying leisure" (Urbanistički zavod Beograda 1972). The plan considered the whole territory of the city, including its constraints and development potential, by applying the principles of integrated planning. Another important contribution was its conclusion that for the whole territory of the administrative area of Belgrade there should be one unified plan, which actually happened in 1981.

The plan confirmed the decision from the previous plan to relocate the main railway station to a new location, Prokop. This opened up the possibility of developing the banks of the Sava (Sava amphitheater), where high and medium density housing was envisaged, as well as business and commercial zones. In order to further develop the concept of the development of the city center, in 1976 a study on the central zone of Belgrade was carried out with projections until 2000. The study clearly defined three centers: the historical town center, New Belgrade, and Zemun, while the main connection between these centers was envisaged in the form of three new bridges and roads which crossed the Sava amphitheater, with the activation of both sides of the Sava's shoreline.

Diverse studies regarding an underground train network in Belgrade were carried out during the twentieth century; however, it still looks like a never-ending story, since today there is still no sign of one being built. The project's development was given to the Directorate for the Building and Reconstruction of Belgrade. Part of the contract was to gather an expert team who would make a complete project proposal. The Directorate formed the *Sector for metro and underground works*. The city of Belgrade Assembly accepted all of the initial information, as well as the first analyses made by the project team in a session held on December 20, 1973. According to the Underground Study, special attention was paid in the Feasibility Study to building the first stage, in which two lines would connect new and old Belgrade and serve the city center (14.2 km with 18 stations). The underground network also included Belgrade's constantly expanding territory, spreading 10 km from the city center.

In the years when the work on the first underground line was supposed to start, major political problems in the former Yugoslavia began, which without any doubt soon influenced all decisions made. This project was stopped just before the first construction phase began. Every new government claim that the underground system is a priority in Belgrade and announces the start of the construction process. Each time, a huge amount of money is given for feasibility studies, but Belgrade still does not have its underground system. The problem of Belgrade's underground is connected to political questions and is always the number one marketing tool during every election campaign (Maletin 1993).

The Urban Master Plan of 1972 had two revisions (in 1985 and 1999) that were made to alleviate the disparities between development and the need to review certain concepts, in order to determine which planning solutions would be more realistic. The plan from 1985 envisaged the intensification of tram traffic in the central zone and toward New Belgrade, which began to put into question the idea of the underground system. Although this is a period of gradually giving up on the metro system, there was no giving up on the detailed planning and construction of pedestrian zones and streets on which the team from the Belgrade Metro Agency also worked. This can be seen as a partial continuation of a European trend, having in mind that the first pedestrian streets in Europe were constructed between 1972 and 1978. The conceptual design was conceived for the pedestrian zone in the city center, the pedestrian boulevard in New Belgrade, and the pedestrian street in Zemun. Belgrade got its first pedestrian street in 1988 on the basis of the regulatory plan for the central zone of Belgrade from 1977 (Vukmirović 2013).

The fundamental reconstruction of Knez Mihailova Street and part of Republic Square was carried out according to the project by architect Branislav Jovin. Work started in May 1988, and the street opened to pedestrians on October 20 of the same year. Converting Knez Mihailova Street into a pedestrian zone was one of the few large urban themes that were brilliantly realized, bearing in mind that Belgrade had not developed enough in the areas of planning and continuity. The Regulatory Plan for the Central zone of Belgrade from 1977 envisaged the formation of a pedestrian zone, i.e., a network of pedestrian streets, but by 1988 only part of the plan had been implemented (Vukmirović 2013). Expansion of the Knez Mihailova Street pedestrian zone continued in 2008, after which 9 more streets were completely or partially pedestrianized (according to the Plan from 1977). This trend continued, but significantly more ambitiously, and in 2015 came the IME project: Identity_Mobility_Environment (Folić and Vukmirović 2015). The project envisages the expansion of the Knez Mihailova Street pedestrian zone and includes the full implementation of the Plan of 1977, which would make a complete entity with the focus on Knez Mihailova Street, as well as continued expansion of the pedestrian streets toward the bank of the Sava on one side and toward Dorćol, that is, the bank of the Danube on the other side.

One focus of the 1985 General urban Plan was the lowering the city to the banks of the Danube in the part that touches the historical core. Expansion of the existing urban matrix to the right bank of the Danube was envisaged, from Kalemegdan to Pančevo Bridge. On the other hand, the shore area of the Sava, known as the Sava amphitheater, was treated in a similar way to the Plan of 1972. This topic became popular again at the end of the '80s and was the theme of two urban architectural competitions for this area: The International competition on the future of New Belgrade from 1985 and the Third Millennium project, organized by the Serbian Academy of Arts and Sciences in 1991. After this period, the Institute for Urbanism carried out a study of *Varoš on the water*, which promotes the idea of Belgrade's descent onto the banks of the river. The study, which treated the left and right banks of the Sava equally and observed them in design and functional terms as a unified whole, placed special emphasis on the area of the Sava amphitheater and the connection with Terazije Terrace. Connection of the left and right banks of the river was envisaged by means of a network of canals and artificial islands, as well as newly planned systems of compact blocks for housing, commercial use, business, tourism, and services. The *Europolis* project from 1995 proposed the construction of *New Manhattan* along the banks of the Sava River. A high degree of development was proposed along the riverbanks, with high-rise buildings on the New Belgrade side and a high index of occupancy on the right bank, with mixed functions on both sides. This is still a current theme, and after these projects, new concepts were offered. However, each proposal was met with similar problems when any further action was taken, those of infrastructural changes and investment.

As can be seen, the first years of the post-war period saw the development of Belgrade as the Yugoslav administrative and cultural center, with a focus on communal and economic and cultural development, which later expanded to housing construction and transport. During this period, certain decisions and actions had a profound influence on the conception of the first post-war General Plan. The most important one was the initiative of the highest state leadership in 1946 to take the city to the left bank of the Sava, building a new part of the city—New Belgrade. Furthermore, residential settlements were built on the peri-urban landscape, such as Karaburma, Kanarevo Brdo, Tošin bunar, and Železnik, while the Belgrade public at that time was most interested in the construction and reconstruction of the central parts of the city. The end of the 1950s saw the creation of a large number of urban plans for the regulation of the central zone, which, unfortunately, were implemented only in fragments. The main reason for this was the economic and political situation, which resulted in almost a sixfold reduction in investments. This is how the idea of "filling up" blocks with new, high-rise buildings on parts of empty plots toward the street came about, while the interior of the plots continued to be occupied with old and low buildings. In addition to this, Tashmajdan Park was built with a sports complex; a new fairground area was constructed on the right bank of the Sava; and Confluence Park and other significant open spaces and public buildings were completed that preceded the Summit of the Non-Aligned Movement held in 1961.

At the beginning of the 1960s, the situation changed, thanks to the first effects of the industrialization of the country, which is reflected in planning and construction. The reorganized Urban Planning Institute, together with the Council for Urbanism, invested great efforts in the rapid growth and transformation of Belgrade. The development followed appropriate policies, measures, and urban plans. By the end of 1974, regulatory plans for the entire territory of the city, urban plans for surrounding suburban settlements, detailed urban plans for New Belgrade, and other large residential settlements and the new General Urban Plan of Belgrade for the period from 1972 to 2000 were developed. Thus, in the period from 1960 to 1974, 127,858 apartments were built in Belgrade (Tasić 1995, 558), which is about four times more than in the post-war period. The general construction of residential settlements (Karaburma, Šumice, Braće Jerković, Konjarnik, Kanarevo brdo, Miljakovac, Kneževac, Gornji Zemun, Donji Zemun, Zemun IV and XI, Dorćol, Vračar, Banjica, Kumodraž, and Kotež) was characterized by the application of Korbizije's urban development concept within the Athens Charter. This was in line with the socialist system and urban planning, especially in terms of the socialist principles of justice, humanity, and equality. In addition, the Ada Ciganlija project was implemented and became the largest sports and recreational center in the city.

After the adoption of the plan in 1972, Belgrade developed at a slower pace, indicating that although the plan was good in terms of strategy and development, it was also too ambitious. However, some important projects were implemented that improved the quality of life of citizens, especially in the city center—the pedestrianization of Knez Mihailova Street, the continued construction of St. Sava Temple, the extension of the city's heating network, the reconstruction of the National Theatre, the arrangement of Republic Square, the renovation of old and valuable façades, etc. Unfortunately, war conflicts during the 1990s and international isolation again broke the developmental continuity and transformation of the city, leaving some strategic sites and big projects (metro, bypass around Belgrade, central train station) waiting for future times.

2.3 Belgrade Urban Plans After 2000

The period after 1990 was characterized by the post-socialist and post-war phase of the breakup of the SFRY. The disintegration of the SFRY was a consequence of a complex set of international circumstances and internal differences that led to political and armed conflicts. It was also due to the "lack of reform of the economy and society, but also the inability of political elites to transform the system" (Zeković and Vujošević 2018, p. 44). Serbia was exposed to the destruction of the economic system, internal cleavage, international isolation, and war devastation, without a clear strategic policy. In 1995, Serbia adopted a "building law package" (strongly criticized by experts) on construction land, on the construction of buildings, on planning and spatial planning, with the aim of creating an environment for attracting foreign investors.

Democratic changes followed after October 5, 2000, followed by a post-socialist development context in which "new institutions are created on the neoliberal capitalist system" (Zeković and Vujošević 2018, p. 45). In 2003, the Law on Planning and Construction (with 25 bylaws) was adopted, within which legal matters related to spatial and urban planning, construction land, and the construction of facilities were unified. The law was amended several times, and all the changes followed the state's efforts to create urban and other legislation that would provide investors with subsidies in the field of construction land, quick and efficient access to cheap and attractive locations, and prompt access to building permits.

The beginning of this phase is related to the adoption of the Master Plan of Belgrade 2021 in 2003 and the Regional Spatial Plan for the Administrative Territory of the City of Belgrade adopted in 2004 and revised in 2009 and 2011. During 2011, Belgrade adopted its first strategy for the development of the city of Belgrade for a period of 4 years. After the expiration of the mentioned period, the next development strategy was adopted in 2017. Following the adoption of the Law on planning and construction in 2009, the framework for the new general plan was created. It was adopted in 2016 and was an improvement of the existing one from 2003, including some modifications.

The Regional Spatial Plan for the Administrative Territory of the City of Belgrade stated as its main goals the protection, spatial planning, and development of the City of Belgrade in correlation with its metropolitan area, based on the principles of sustainable development. The plan defined the basic vision of the development of Belgrade that seeks to elevate the City of Belgrade to a high-ranking level among the metropolitan cities and the capitals of Central, Eastern, and Southeastern Europe, according to the criteria of a sustainable economy and advanced technology.

The City of Belgrade Development Strategy was adopted in March 2011 with the basic goal "to find the best way to raise the City of Belgrade to a higher level, the level of big European cities" (Palgo Center 2011). The City of Belgrade's special role is the task of strengthening its position in the constellation of European capitals, but also enhancing economic and cultural structures, and raising the ecological and aesthetic standards of urban and rural areas to a higher level. The new City of Belgrade Development Strategy 2021 was adopted in 2017. The authors of the strategy, RAREI and the Palgo Center, with a team including a board of experts and members of the local government, state the vision of the City of Belgrade as an "urban, sustainable, responsible, dynamic and developed European region" (RAREI and Palgo Center 2017). The strategy defines eight goals that correspond with established development sectors. The city's identity is marked mainly by its central zone, and in absolute terms, its position on the mouth of Sava and the Danube. According to the City of Belgrade Development Strategy, in the future, emphasis will be on restoration of the existing urban structures, especially the central parts. Natural and cultural heritage will have a special role in creating Belgrade's new identity.

Development of the new General Plan of Belgrade 2021 took place in a society in transition and was adopted in September 2003 (Fig. 2.11). Having lost control over its own development during the '90s, only to reach a low level of daily functioning after 2000, Belgrade needed a plan of "reparation, regeneration, renovation and reconstruction, i.e. a plan for great correction" (Bojanić 2013). Great correction includes almost everything: abandoned industrial zones, a bad transport system, spontaneously developed economic activity and trade carried out in buildings that have usurped public areas, poor city areas, and illegal construction, significant buildings destroyed during the bombing, etc. The culmination of

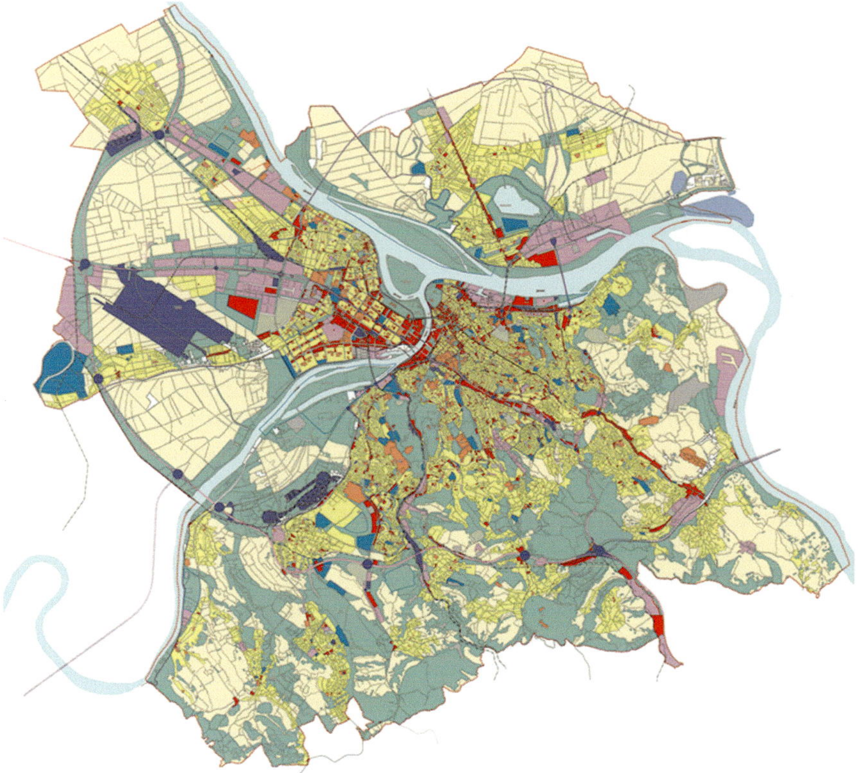

Fig. 2.11 General urban plan of Belgrade 2020 from 2003. Archive: Belgrade Urban Planning Institute

informal building in Belgrade came in the 1990s as a result of the decades of devastation and destruction of the city. According to the Master Plan 2021, around 35,000–40,000 illegal buildings were built in Belgrade in 1994 alone, which corresponded to about 100,000 inhabitants (GUP 2021). Out of a total number of 4.6 million buildings in Serbia in 2014, 28% (1.3 million) were informal structures. The plan sets 12 main objectives, including the capital of the country, European metropolis, orientation toward the Danube, harmony with nature, sustainability, the regulation of urban planning, economic vitality, connected and accessible, the city of culture, and others. Also, the long-term concept of spatial organization and development of Belgrade is through large-scale projects.

The dual purposes of this plan, on one hand to make major repairs, and on the other hand to assess and open development opportunities, have shaped the very character of the document, which is characterized by flexibility rather than rigidity. It is dynamic rather than static and supports the processes rather than just being a "picture." The approach to the separate General Urban Plan that involves continuous planning and improvement highlights "an annual approach to amendments," and it has already resulted in the adoption of four amendments, with another currently being adopted. This last amendment improves segments of the planning solutions related to agricultural areas, religious buildings, tall buildings, the Belgrade shoreline, environmental protection, renewable energy sources, and the energy efficiency of buildings. However, the backbone of the planned development of Belgrade considers the identification of potential sites for large city projects as areas of special interest for the city, the rehabilitation and transformation of old industrial and military complexes, and previously defined planning solutions for economic zones and parks.

These large city projects of particular interest are focused on areas of urban transformation and regeneration, abandoned industrial and military complexes, brownfield locations, and the Belgrade shoreline. They also include new housing construction projects, affordable and social housing, strategic transport projects, and affirmation of the urban identity and values of the city area.

In a similar way to the first versions of this plan, the GUP Belgrade 2021 and its amendments offer new interpretations of old ideas: the metro, collective housing, reappropriation of space, treatment of the greenery in the city, and transformation of the highway. However, what has finally come to order, which runs through all of these plans, is the regulation of the Sava and Danube shoreline. Until recently, the riverfront has been characterized by devastated industrial zones, with public promenades in poor condition; however, several megaprojects may completely change this situation. The most ambitious ongoing flagship project is Belgrade Waterfront (more in the chapter on Belgrade's Waterfront redevelopment). The main objective of the latest version of Belgrade's General Urban Plan (2016) is transformation of the urban planning system in accordance with market-based socioeconomic, political, institutional, and organizational changes (Zeković and Vujošević 2018, pp. 49–50). The direct impact of market and investor interest is evident in the megaprojects mentioned. The strategic objective of land management is to establish a new management model based on market principles and to correct market imperfections, with the support of the public interest. Urban reconstruction is planned, in parallel with an increase in built construction land by 50%.

The transition process began after democratic changes in 2000. The period that led up to this was one of war and bombardment, international isolation and sanctions, and economic collapse which contributed to Serbia being one of the most underdeveloped countries in Europe, with modest development prospects. This period is characterized by poorly developed institutional and democratic culture, and an underdeveloped market in which urban planning becomes a partner. In such circumstances, several megaprojects were announced, of which, Belgrade Waterfront, is currently under construction, while the potential of the other projects is being manipulated in order to find investors and launch their implementation. The extension of the pedestrian zone is also underway, which has an analogy in relation to the Central Zone plan from the 1970s. Most of these processes are nontransparent, while public discussions are undermined and targeted at less relevant goals and purposes. An uncertain period is ahead of us, and we will be able to evaluate it only in the time ahead.

2.4 Conclusion

The plans and development concepts of Belgrade presented here with reviews on the level of realization, as initially stated, can point to the ambitions and aspirations of their initiators and creators, as well as to the social and political circumstances that enabled or prevented the realization of the plans. Although defined by major crisis events in the world, the separate periods that have served here as a framework for presenting the basic ideas of the plans are also determined by specific sociopolitical circumstances in which the state and its capital were then situated.

The character of the turbulent moments can be traced from the fact (Jovanović 2004, 198–199) that Belgrade was bombarded and demolished five times (1914, 1915, 1941, 1944, 1999) and that it was the capital of six states 1830–1878; 1878–1918; 1918–1944/45; 1945–1991; 1991–2003; 2003), which fought seven times (1875–1878, 1885, 1912–1913, 1913, 1914–1918, 1941–1945, 1991–1999), in which five radical changes of dynasties and rulers occurred (1903, 1941, 1945; 1987/88, 2000), followed by six more or less radical transformations of the social organization, and the rapid destruction of the old and the establishment of new political, state, and economic structures (1903, 1918, 1944/45, 1948, 1988/91, 2000) in which there were nine radical shifts in foreign policy (1877/78; 1903; 1917/18; 1941; 1944/45; 1948; 1957/61; 1987/88; 2000).

The first frame of Belgrade's urban development was characterized by a period of liberation within which it turned to Europeanization and endeavored to eliminate Oriental elements from its urban fabric. During this period, there was a change of monarchy and the emergence of the first ambitions for the formation of a larger state. This happened after the First World War, which also determined the second period of Belgrade's urban development. The first general urban plan was adopted, and Belgrade recorded its highest level of construction in a very short period of time. This was interrupted by the Second World War, during which the capital was twice bombarded. The third period was characterized by the establishment of new sociopolitical structures and a self-governing approach, and over a period of almost four decades, the state was involved in various international partnerships including the USSR, Europe, and the Non-Aligned Movement. At the same time, the state became economically stronger, thus generating new ideas that would shape the capital. During the 1970s, Belgrade reached the culmination of planning, after which everything became weaker under the new circumstances that announced the beginning of the transition processes and the crisis in the territory of the former Yugoslavia. International isolation and another break in the development of urban continuity led to the fact that Belgrade entered the new millennium with unclear strategic definitions and a plan designed to solve problems, rather than properly confronting the new challenges of liberal capitalism embedded in the image of the city.

According to the above, the city and its planning documents can serve as an important document of its life and aspirations, which "helps in a clear reading of political intentions and concepts inspired by individual urban procedures." Therefore, in the following chapters, we will try to present various topics that stand out in Belgrade, and which will determine its appearance and development in the times ahead.

References

Arandelovic B, Vukmirovic M, Samardžic N (2017) Belgrade: imaging the future and creating a European metropolis. Cities 63(2017):1–19

Blagojević L (2009) Urban regularisation of Belgrade, 1867: trace vs. erasure. SAJ 2009-1

Bojanić B (2013) Intervju sa Vladimirom Macurom - kristalizacija ideja kao razvojna perspektiva Beograda. Arhitektura i urbanizam 37:57–71 (Interview with Vladimir Macura—crystallization of ideas as a development perspective of Belgrade)

Djordjević D, Dabović T (2010) Ideology and practice of town planning of Belgrade 1867–1972: the rise in collection of papers, vol LVIII

Folić M, Vukmirović M (2015) Projekat IME: Identitet_Mobilnost_Ekologija Grada Beograda. Grad Beograd (Project IME: Identity_Mobility_Enviromnent)

Gašić R (2011) Problemi teritorijalnog širenja Beograda između dva svetska rata in Istorija 20. veka 3/2010, Beograd, pp 57–68 (Problems of the territorial extension of Belgrade between two world wars)

Ilijevski A (2013) Prilog proučavanju arhitekture i ideologije mosta viteškog kralja Aleksandra I Ujedinitelja, Contribution to the study of architecture and ideology of the Bridge of King Aleksander I Unifier in Belgrade, Heritage, No 14, pp 211–220

Josimović E (1997) Objasnenje predloga za regulisanje onog dela varoši Beograda što leži u Šancu. Beograd. Društvo urbanista (Explanation of the proposal for regulation of that part of the city of Belgrade that lies in moat)

Jovanović M (2004) 200 Godina Moderne Srbije ili Dva Veka Neravnomernog i Prekidanog Istorijskog Razvoja in Godišnjak grada Beograda LI pp 193–212. (200 years of Modern Serbia and two centuries of the uneven and interrupted historical development)

Lazarević-Radak S (2011) Urbanizacija i karakter grada. Oblikovanje grada u delima britanskih putnika: Primer Beograda 1926–1939. godine In Prostorno planiranje u Jugoističnoj Evropi (do Drugog svetskog rata), edited by Miljković-Katić B Beograd: Istorijski inistitut/Balkanološki inititut SANU, pp 229–250 (Urbanization and the character of the city. The design of the city in the works of British travelers: example of Belgrade 1926–1939)

Maksimović B (1967) Emilijan Josimovic - prvi srpski urbanist. IAUS: Beograd (Emilian Josimovic—the first Serbian urbanist)

Maksimović B (1983) Ideje i stvarnost urbanizma Beograda: 1830–1941, Zavod za zaštitu spomenika kulture grada, Beograd (Ideas and the reality of the Belgrade's Urbanism)

Maletin M (1993) Metro kao osnovo urbanističkog razvoja Beograda, publikacija Put i saobraćaj, Beograd, pp 18–27 (Metro as the basis of urban development of Belgrade)

Medaković D (1997) O Emilijanu Josimovicu (izabrani tekstovi). Društvo urbanista, Beograd (About Emilijan Josimovic)

Obradović J (1932) Primena Građevinskog zakona na uređenje i izgrađivanje Beograda in *BON* (BON) 11 (1932): 695 (Application of the Building Law to the regulation and construction of Belgrade)

Obradović J (ed) (1938) Građevinski zakon. Planeta, Beograd (Building Law)
Perović M, Krunić S (eds) (1998) Nikola Dobrović - eseji, projekti, kritike. Beograd. Univerziteta u Beogradu - Arhitektonski fakultet (Nikola Dobrović—essays, projects, criticism)
Petričić B (1976) Prve urbanisticke realizacije: New Belgrade 1955–1975, Beograd (The first urban realizations: New Belgrade 1955–1975)
Perović M (2003) Istorija srpske arhitekture XX veka. Od istoricizma do drugog modernizma (Serbian 20th century architecture. From historicism to second modernism), Beograd (Belgrade) Univerzitet u Beogradu - Arhitektonski fakultet (University of Belgrade - Faculty of Architecture)
Rastović A (2013) Britanska percepcija Beograda: Modernizacija u XIX veku in Limes PLUS Stranci u Beogradu, no 2 pp 27–39 (British perception of Belgrade: modernization in XIX century)
Ristović M, Stojanović D, Jovanović M, Perišić M, Miloradović G (2008) Živeti u Beogradu 1890–1940. Dokumenti Uprave Grada Beograda. Beograd: Istorijski arhiv Beograda (Living in Belgrade 1890–1940. Documents of the Administration of the City of Belgrade)
Roter-Blagojević M, Radivojević A (2007) Les espaces publics et la vie publique a Belgrade au XVIII et au XIX siecle et leur transformation au XX siecle in: Etudes balkaniques: recherches interdisciplinaires sur les mondes hellenique et balkanique Cahiers Piere Balon 14. L'espace public dans le Sud-Est europeen (edit. prof. Andre Guillon). De Boccard. Paris: 105–142
Stojanović D. Grad i politika Published online in Peščanik on July 07, 2015. https://pescanik.net/grad-i-politika/. Accessed 10 April 2019 (City and politics)
Stojanović B, Martinović U (1978) Beograd: 1945–1975: arhitektura i urbanizam. Tehnicka knjiga, Beograd (Belgrade: 1945–1975: architecture and urbanism)
Tasić N (1995) Istorija Beograda. Balkanološki institut SANU, Beograd (The history of Belgrade)
Terzić D (1997) Emilijan Josimović, vizionar ulaska Beograda u urbanu civilzaciju, Društvo urbanista Srbije, Beograd (Emilijan Josimovic, visionary of Belgrade's entry into urban civilization)
Urbanistički zavod Beograda 2000 (1972) Generalni urbanistički plan Beograda. Beograd. Urbanistiki zavod Beograda (general urban plan of Belgrade 2000)
Vujošević M, Nedović-Budić Z (2006) Planning and societal context—the case of Belgrade, Serbia. In: Tsenkova S, Nedović-Budić Z (eds) The urban mosaic of post-socialist Europe. Contributions to economics. Physica-Verlag HD, Heidelberg, pp 275–294
Vukmirović M (2013) Značaj i uloga mreže pešačkih prostora u generisanju kompetitivnog identiteta grada. Doktorska disertacija. Univerzitet u Beogradu - Arhitektonski fakultet (The importance and role of the pedestrian space network in generating the competitive identity of the city)
Vukotić-Lazar M, Djokić J (2006) Complex history as a source of planning problems: old Belgrade fairground. Spatium (13–14):34–40
Vuksanović-Macura Z (2014) Singidunum Novissima: Novo čitanje poznatog konkursnog rada in Nasleđe. no 15, pp 115–127 (Singidunum Novissima: a new read of the famous competition work)
Vusanović Macura Z (2014) Generalni plan Beograda 1923: Komparacija planiranog i ostvarenog. Vol. Doctoral thesis, University of Belgrade - Faculty of Architecture, Belgrade (General plan of Belgrade 1923: Comparison of the planned and realized)
Zeković S, Vujošević M (2018) Construction land and urban development policy in Serbia: impact of key contextual factors. In: Bolay JC, Maričić T, Zeković S (eds) A support to urban development process. EPFL, IAUS, Lausanne/Belgrade, pp 29–58
Zeković S, Maričić T, Vujošević M (2016) Megaprojects as and instrument of urban planning and development: example of Belgrade waterfront. In: Hostettler S, Najih Besson S, Bolay JC (eds.) Technologies for development UNESCO 2016. Springer, Cham, pp 153–164

Chapter 3
The Rise and Fall of New Belgrade with Current Developments in Retrospect

Biljana Arandelovic

Abstract New Belgrade is a modern city within today's Belgrade metropolis, built in the second half of the twentieth century in the sand-filled flood plain of Bežanija Field, located on the left bank of the River Sava and right bank of the River Danube between the historical cities of Belgrade and Zemun. The terrain where New Belgrade was developed was for centuries a no man's land between the Ottoman and Austrian Empires. The construction of New Belgrade after WWII was a major urban project in the former Yugoslavia, which resulted in an influx of new residents from within Serbia. The initial concept of New Belgrade was as a modern, governmental, and central city in Yugoslavia, with the main function of being the seat of power of the new federal administrative structure. However, New Belgrade was completed as a city with an utterly different function; it was predominantly a city where people lived, and not a seat of power or a primary aesthetic object of the state, as initially planned. The monumental architecture of New Belgrade never fulfilled the role of representing a governmental city, as it was completed with a delay, and in different architectural and town planning circumstances than those initially planned. The intention to equalize the central zone of New Belgrade with the historical center of Belgrade did not succeed at first. The center of New Belgrade has remained until recently an ugly field of housing blocks. This part of the city is still subject to change. This chapter features the most important structures in New Belgrade. For example, Hotel Yugoslavia which was one of the first three significant facilities planned in New Belgrade, next to the Central Committee building and the Presidential Government building, today the Palace of Serbia. The others are the Museum of Contemporary Art, one of Belgrade's architectural masterpieces and Sava Convention Center, an iconic postmodernist New Belgrade complex which was a Pritzker Prize nomination in 1979. The chapter also presents Blok 12, which has become the base for most successful Serbian start-ups, and also some of the current urban development projects such as the construction of the Belgrade Philharmonic within Block 13, the relocation of the bus and train station from the old center in Belgrade to New Belgrade Block 42 and the Block 18 development project.

Keywords New Belgrade · Yugoslavia · Socialist city · Post-socialist city · New Belgrade urban development · Urban transformation

© Springer Nature Switzerland AG 2020

B. Arandelovic and M. Vukmirovic, *Belgrade*, The Urban Book Series,
https://doi.org/10.1007/978-3-030-35070-3_3

3.1 New Belgrade Introduction

New Belgrade (Novi Beograd) is a modern city within today's metropolis of Belgrade, built in the second half of the twentieth century in the sand-filled flood terrain of Bežanija Field (Bežanijsko Polje). The Municipality of New Belgrade was officially formed on May 30, 1952, and today it is the largest of the city's 17 municipalities. The construction of New Belgrade after World War II was a major urban project in the former Yugoslavia. The initial concept was for it to be a modern, governmental, and central city in Yugoslavia, with the main function of being the seat of power for the new federal administrative structure.

However, New Belgrade was completed with an utterly different function; it was built predominantly as a residential municipality, and not a seat of power or a primary aesthetic object of the state, as initially planned. The monumental architecture of New Belgrade never fulfilled the role of housing the government, since its construction was delayed and took place in different architectural and town planning circumstances than those initially planned.

The intention to equalize the central zone of New Belgrade with the historical center of Belgrade did not succeed at first. New Belgrade started very ambitiously and for a long time was a huge dormitory—until the 1990s. During the 1990s, illegal construction was the dominant phenomenon in the urban development of the city, and this affected New Belgrade in particular as it had a lot of empty space. Many of the earlier residential buildings constructed in New Belgrade were designed to have common spaces in them that belong to all of the tenants. These spaces were mostly used for storage, laundry, or as a gathering and socializing room. During the 1990s, most of them were suspiciously privatized and converted into apartments or office buildings. Numerous ground floor levels of residential buildings in New Belgrade's blocks were adapted to become stores, while kiosks were built every few steps almost everywhere: on sidewalks, between residential buildings, on the riverbank, even in parks and green areas, as temporary facilities. The kiosks were mostly metal constructions operating with temporary licenses, although they became illegally replaced with permanent constructions made of bricks. Many kiosks are still there and have become an integral part of the urban fabric of New Belgrade (Figs. 3.1 and 3.2) (Jocić 2019).

Since the 1990s, New Belgrade has been radically transformed under a market economy, and it has become fully open for foreign investments, resulting in constant planning activities. It has been possible to witness its unique transformation into Belgrade's so-called "Manhattan" (Vujović and Petrović 2007). This part of the city is still subject to change. New Belgrade has again become a massive construction site, but now the new period of globalization and international investments has brought modern high-standard shopping malls and the headquarters of multinational companies. While the old city center is overbuilt, New Belgrade has enough available plots for construction investments. Buildings in the old city center cannot meet the needs of modern businesses or offer the conditions for new offices, but New Belgrade can.

Fig. 3.1 Street trade Kiosk, New Belgrade. Photographer Biljana Arandelovic. Rights reserved

Fig. 3.2 View of the shops located outside the residential buildings in New Belgrade. Small-scale trade facilities. Photographer Biljana Arandelovic. Rights reserved

The transport infrastructure is better organized and developed in this area, with wide boulevards and modern roads—unlike the central district in the old town, where the inherited narrow street network is overcrowded (Jović and Djorić 2010). The transport connection is outstanding through New Belgrade and it connects with the whole of Europe through Corridor 10. Belgrade's Nikola Tesla Airport borders the edge of the New Belgrade Municipality, which is a great advantage for the development of business. The only drawback is New Belgrade's weaker connection to the old city center, despite the construction of the new Ada Bridge (Jovanović and Ratkaj 2014). This all demonstrates clear advantages for new business investment over the old city center. It is also important to emphasize New Belgrade's particular spatial position within the city of Belgrade, located just a few hundred meters away from the old city center, on the other side of the rivers.

New Belgrade today has around 200 skyscrapers, with 72 apartment blocks (Figs. 3.3, 3.4, and 3.5) and six hundred large residential buildings. Family houses can be found only in the Bežanijska Kosa and Ledine districts, both of which were built as temporarily settlements in the 60s, which then became permanent. The remaining 80,000 apartments in New Belgrade are located in multi-storey apartment blocks. With numerous alleys, parks, and wooded areas, New Belgrade is the municipality with the most greenery in Belgrade (novibeograd.rs). It covers an area of 41 km^2, with a total population of 214,132 according to 2017 statistics (Statistical Office of the Republic of Serbia 2017).

3.2 New Belgrade: From No Man's Land to a Modern City

The terrain where New Belgrade is situated had for centuries been a no man's land between the Ottoman and Austrian empires, with the function of a border and sanitary cordon; it was a center of warfare and a place of historical changes during the twentieth century. The terrain is located on the left bank of the Sava and right bank of the Danube, between the old historical cities of Belgrade and Zemun. In the Roman period, this land was situated at the northern edge of the Empire, while after 395 A.D. it was exactly the area that separated the Eastern and Western Roman Empires. Thus, the future territory of New Belgrade was for most of its history separated from the historical core of Belgrade on the other side of the Sava River. After the evacuation of the Turks from Central Europe at the end of the eighteenth century, this land became a tampon zone between the Ottoman and Habsburg Empires and later, after the Serbian wars of independence in the nineteenth century it became a no man's land between Serbia and Austria-Hungary. It was not until 1918, after formation of the Kingdom of Serbs, Croats, and Slovenes, that the territory of what would later be New Belgrade and the old city core of Belgrade became one united city (Kulić 2014).

The journey of New Belgrade's urban development is long and turbulent, and its direction has changed many times, always in connection with political power. It is still subject to change but is no longer just *a big dormitory*, which is how it was described for years. The construction of New Belgrade after World War II was a major urban project in the former Yugoslavia, which resulted in an influx of new residents from within Serbia.

However, the first idea about expanding Belgrade to the left bank of the Sava River appeared much earlier, and it is closely connected with the construction of Belgrade fairground, which has a special and significant place in the town planning

Fig. 3.3 Apartment blocks New Belgrade. Photographer Biljana Arandelovic. Rights reserved

Fig. 3.4 Apartment blocks New Belgrade. Photographer Biljana Arandelovic. Rights reserved

development of Belgrade. The fairground's construction drew attention to the new prospects for town planning in Belgrade, with broader ideas than those found in the General Plan of 1923.

Apart from the economic importance of Belgrade fairground, it was a signal to continue development on the left bank of the Sava. This was further encouraged by the newly established highway and tram line for Zemun that commenced operation in 1931. Another important year from this period was 1937, when a Danish business corporation offered to fill a section of the land on the left bank of the Sava with sand, so that this space could be used to expand Belgrade. They also submitted a town plan for the future settlement. The entire issue of Belgrade crossing over onto the left bank was considered a very profitable business, both for the Danes and for Belgrade (Maksimović 1983; Ristanović 2009).

By the end of 1939, there was, most likely, a disagreement in the negotiations, or some other unforeseen difficulties, so the Danish group started selling the sand-filled area to contractors. Everything was interrupted by the fascist attack on Yugoslavia. The first phase of New Belgrade's development was the extensive housing construction which began in the post-war period. New Belgrade was an integral part of the first post-war General city plan of Belgrade, adopted in 1950.

On January 1, 1947, the call for big competitions in New Belgrade was announced, namely, the public Yugoslav competitions for the preliminary design of

Fig. 3.5 Apartment blocks New Belgrade. Photographer Biljana Arandelovic. Rights reserved

the Central Committee of the Communist Party of Yugoslavia (CK KPJ) and the Presidential Government building of the Federal People's Republic of Yugoslavia (FPRY, then the Federal Executive Council—SIV). This also included projects, in addition to individual facilities and proposals, for the urban plan of New Belgrade as a new governmental city of the Federal Republic of Yugoslavia. Besides these two major competitions, a third was announced for a modern hotel to represent the state (Arandelovic et al. 2017; Blagojević 2005, 2007; Macura 1947).

The New Belgrade zone was divided by its main traffic routes into four longitudinal belts; the one closest to the riverbank was intended for administrative and governmental institutions, and the remaining three for housing. In terms of its traffic solution, the plan was based on an orthogonal system adapted to the curves of the rivers, with longitudinal traffic routes of the first order, and with the main highway in the middle of the New Belgrade plan (Maksimović 1983).

In 1948, the first work on public utilities in Bežanija Field began. Soon after, the construction of the first structures started: the presidential government building of the FNRJ government, a modern hotel, student halls of residence and the pavilions at the Tošin Bunar location (Blagojević 2007). These works were canceled after just 2 years because of the political and economic crisis caused after the break with Soviet communism in 1948.

The initial concept and idea of New Belgrade was as the modern seat of power of the new federal administrative structure for the FNRJ. However, it took on an utterly different function, being predominantly a Belgrade municipality where people lived, and neither the seat of power nor a primary aesthetic object of the state as initially planned. The monumental architecture of New Belgrade never fulfilled this role, as construction did not really begin until the 1960s in totally different architectural and town planning circumstances than initially planned. The first physical structures in the deserted landscape of Bežanija Field, which contrary to the ambitious vision of New Belgrade actually represented the new socialist architecture, were the housing blocks. The basis of the housing policy, which was put in practice in the building of New Belgrade and which was subject to the collective ownership concept in socialism, was the ideological assumption that an apartment was an asset of the society, just as, for example, city greenery or infrastructure (Blagojević 2005; Djurović 1995; Kulić et al. 2012).

The construction of housing units was planned in a centralized way, with a desire to build many flats and solve the burning housing issue. The consequence of this was that in the first 10 years after the war, the housing standard significantly decreased compared to before the war. The population increased by 160,000, which required the construction of 35,000 new flats, but by 1956, only around 14,500 had been constructed. With a number of households living in shared flats, there were a total of around 57,000 households more than the number of available flats. By the mid-1950s, privately owned housing still prevailed in Belgrade (around 79%), while 89% of housing built after the war was owned by the state. The structure of newly built flats was modest, since the shortage of small flats inherited from the previous period had to be addressed. This is the reason why mostly two-roomed flats were built (Stojanović and Martinović 1978).

The funds for housing construction were formed by obligatory tax on the gross income of the working population, and from common funds. By the middle of the 1960s, the finances for these purposes were centralized into special, so-called housing funds. After switching over to the communal system, the source of funding remained the same, but the main social investors were now the City of Belgrade, Yugoslav People's Army and State Secretariat of Foreign affairs, instead of the Federation or the Republic. Apart from that, a large number of flats in New Belgrade were directly purchased from the budget resources of the organs and organizations of the Federation, Republic, city, and social organizations. Construction was on one hand dictated by ambitious and unrealistic 5-year plans, and on the other hand hindered by poverty and a lack of modern construction technology. The results were poor aesthetic and urban qualities (Maksimović 1983; Blagojević 2005).

Until the beginning of the '60s, newcomers to New Belgrade lived in pavilions, barracks, and student hostels at Tošin Bunar, entirely isolated from the historical part of Belgrade by the deserted Bežanija Field, and they were oriented toward the central functions of the historical part of Zemun. During this period of stagnation in the construction of New Belgrade, a wave of illicit construction occurred.

The housing plan envisaged strictly defined construction types, frontal tracts in a row, and a system of freestanding structures in mega-blocks (500 × 650/750 m) in the central zone, or in smaller housing blocks (250 × 300 m) following the geometry of the left bank of the Sava behind the railway station (Maksimović 1983).

The housing zone was arranged in three areas with a total of 80,000 inhabitants, and a density of 300 citizens per hectare. Groups of housing blocks 4–8 storeys high, with services in the ground floor, were conceived as a system of freestanding pavilions of parallel structures with greenery. Housing blocks 1 and 2, with around 3,600 flats on a surface area of 35 ha, were designed between 1958 and 1959 according to the provisions of the General City Plan of New Belgrade, and then constructed 1959–1963.

Considering the immense mechanical increase in Belgrade's population after the war, when the city population was young, there was an abrupt growth in the birth rate, which culminated in 16.5/1000 in the '50s which was the highest recorded in the twentieth century until then. These demographic changes, as well as the depleted housing capacity, caused a severe shortage of flats in the years following the war. From 1945, when the average index of useful housing surface area per citizen was 12 m^2, it dropped to 10.5 m^2 per citizen in 1960. In Belgrade in 1955, more than 50% of flats were smaller than 35 m^2 (Djurović 1995).

The intention to equalize the central zone of New Belgrade with the historical center of Belgrade did not succeed at first. For a long period, no trade, business, or cultural structures were built in New Belgrade, and until recently it remained an ugly field of housing blocks for a specific social structure of inhabitants, who lived in modest two-roomed flats of socialist standards.

3.3 Palace of Serbia

The first prize in a 1947 competition for the Presidential Government building of the Federal People's Republic of Yugoslavia (FPRY, then the Federal Executive Council—SIV) was given to a team of architects from Zagreb made up of Anton Ulrich, Vladimir Potočnjak, Zlatko Najman, and Dragica Perak. The second project of importance to the state, announced at the same time, was for the Communist Party of Yugoslavia's Central Committee buildings. Both projects aimed to demonstrate the power of the new state by means of new representative architectural directions (Blagojević 2005). Additionally, according to the General Plan of 1950, two new facilities for the public were planned to be built in Block 13: the Museum of the Revolution of nations and nationalities of Yugoslavia and the new building for Radio-Television Belgrade. However, this never happened.

Construction on the presidential building commenced almost as soon as the project winners were announced, and almost the entire reinforced concrete structural frame was established. The project had some similarities with Moscow's Centrosoyuz building designed in 1928 by Le Corbusier. These similarities can be seen in the H-shaped base and side entrances. Not long after construction had begun, however, in the middle of new political circumstances, came the decision to stop the construction work.

What began as a presidential Palace for the Yugoslav Government, due to changes in the organizational system of government buildings, was completed as the Federal Executive Council building in 1961. After the political break, construction continued 6 years later, in 1954, under eminent Belgrade architect Mihailo Janković. He was one of the most important representatives of the Serbian and Yugoslav post-war architectural scene, whose creative work has made a great contribution to the construction of Belgrade. He is also the architect of 25 May Museum, built in 1962 with the purpose of hosting all the gifts Tito received during his leadership and as the Central Committee headquarters of the Communist Party of Yugoslavia. The original project was changed and modified in accordance with the new tendencies in architecture, by accepting the principles of the new monumentalism of International style. The architect had to remove any traces of Socialist Realism. The building was officially opened in September 1961 on the occasion of the First Conference of Heads of State and Government of the Non-Aligned Countries (Kulić 2013).

The ambitiousness of the government with this project can be seen in the total building area of 64,235 m^2. At the peak of political power in Yugoslavia over 8,000 employees in the state administration were working in this building (Mišić 2011), and it became the central motif and reference point for the urban concept of New Belgrade. The material emphasized the monumentality of the building, as the entire facade was covered in white marble from Brač. The palace is symmetrical, with a central core and two side wings (Fig. 3.5). It includes six lounges, one for each of the six republics of the former Yugoslavia, as well as thousands of square meters of office space.

Each lounge was carefully designed to include the traditional motifs of its republic, including wall decorations, mosaics, sculptures, and furniture. The building's main lounge, the "Yugoslav salon," is covered with monumental fresco paintings and mosaics (it can accommodate around two thousand people). Every part of the interior is special but the most impressive element is the skylight chandelier designed to resemble the sun (18 m in diameter weighing nine tons). The interior is covered in marble from all over the country, selected from each region. These individual entities represent a kind of micro-museum of the nations and nationalities of Yugoslavia: Macedonian, Montenegrin, Serbian, Bosnian, Croatian, and Slovenian. The special value of this building is seen in the artwork collection of Yugoslav painting, sculpture, and works of applied art of the twentieth century, all of which are still in the building, as permanent collections. The palace is filled with works of art by the most eminent artists of the time, and it is a symbol of an era.

Since the 1990s, this building has been the seat of the Federal Republic of Yugoslavia (FRY), then the seat of the Council of Ministers of the State Union of Serbia and Montenegro and since 2006 the Government of Serbia. As the country has changed, so has the name of the building, though most people know it as SIV. In 2006, after Montenegro left the State Union, it was renamed the Palace of Serbia (Figs. 3.6, 3.7, 3.8, 3.9, and 3.10). The Serbian government declared SIV Palace a cultural monument in 2013.

Fig. 3.6 SIV Palace—the Federal Executive Council in New Belgrade. Palace of Serbia in 2010. Photographer Biljana Arandelovic. Rights reserved

Fig. 3.7 SIV Palace—the Federal Executive Council in New Belgrade. Palace of Serbia in 2010. Photographer Biljana Arandelovic. Rights reserved

SIV Palace is located in Block 13 in the area between Mihajlo Pupin and Nikola Tesla Boulevards. According to the current urban development in the area of Block 13, the Philharmonic building will be built next to SIV Palace, based on the winning solution of Dragan Marčetić and Milan Maksimović from a 2016 competition.

3.4 Ušće Towers Within Block 16

Another building that is a symbol of an era is the former Central Committee building (CK), today Ušće. The building was planned at the same time as SIV Palace (the Presidential Government building of the Federal People's Republic of Yugoslavia), and the federal competition for its construction was organized in the same year. Alongside those two projects, a third competition was announced for a luxury hotel for state guests.

The competition for the Central Committee building underlined the importance of making a construction that would clearly show the power of the state, and architects were expected to demonstrate the superiority of socialist architecture. The building was supposed to be dominant in the urban composition of New Belgrade.

Fig. 3.8 SIV Palace—the Federal Executive Council in New Belgrade. Palace of Serbia in 2018. Photographer Biljana Arandelovic. Rights reserved

Fig. 3.9 SIV Palace—the Federal Executive Council in New Belgrade. Palace of Serbia in 2019. Photographer Biljana Arandelovic. Rights reserved

Fig. 3.10 SIV Palace—the Federal Executive Council in New Belgrade. Palace of Serbia in 2019. Photographer Biljana Arandelovic. Rights reserved

The main award went to a project by Slovenian architect Edvard Ravnikar, but unlike the Federal Executive Council building, the project was never constructed.

Eventually, the Government gave up this project and another competition was organized in 1960, which was won by Mihailo Janković, Dušan Milenković, and Mirjana Marijanović. The building was completed to serve a larger number of sociopolitical organizations. It was accordingly named the Building of Social–Political Organizations and was built as a 24-storey tower.

The rectangular tower, with a curtain wall facade of aluminum and glass, known also as “an American facade,” resembles the Seagram Building by Ludwig Mies van der Rohe in New York from 1958. The building was constructed in the period 1961–1965 in a rectangular form as a simple aluminum and glass administrative-type building in International style. The irony is in the fact that the seat of the communist party took a similar form to that of corporate capitalist buildings. Until the end of the twentieth century, this building was the most powerful location in the country, where the party resolved all the important questions of government, the seat of power (Kulić 2013).

After the collapse of the Communist Party of Yugoslavia, the largest part of the building was appropriated by the Socialist Party of Serbia (SPS), whose leader was Slobodan Milošević. The Socialist Party of Serbia declared itself the only legitimate successor to the Communist Party. Most of the available space in the building was rented out, which at one point brought large profits to the new Socialist Party. There was much debate about who the building actually belonged to, and who had the right to profit from the lease, which also included government representatives from former Yugoslav republics. The building was damaged in the NATO bombing of 1999. This building has a greater symbolic value than it does architectural value, as a decision-making spot in the former Yugoslavia.

The former Central Committee building (CK) (Fig. 3.11) was sold in 2002 to a Serbian businessman, owner of MPC Holding and MPC Properties, Petar Matić, who reconstructed it into business office space. The iconic former Central Committee building (CK) was subsequently renamed Ušće Tower, after its location in Block 16—at the confluence of the Sava and Danube (“ušće” means confluence). The reconstruction lasted from 2002 until 2005, during which the building was fully modernized and adapted to the highest standards, receiving two additional floors.

As part of the new urban development of this area, Ušće shopping mall was built in 2009 (Figs. 3.12 and 3.13) in the immediate surroundings. An international competition to find an urban design solution for the Ušće multifunctional business and commercial complex in Block 16 was announced in February 2003, for the space between Nikola Tesla and Mihajlo Pupin Boulevards and Ušće Street. The first prize project by architect Branislav Redžić’s team predicted two towers and a shopping center.

In 2018, the foundation stone was laid for the second tower in Ušće business center, which will have 22 floors. The investor and owner, MPC Properties, is the same as for Ušće Tower. This building, with a gross area of 28,000 m^2 (of which

Fig. 3.11 Ušće Tower today. Photographer Biljana Arandelovic. Rights reserved

Fig. 3.12 Ušće Shopping center. Photographer Biljana Arandelovic. Rights reserved

Fig. 3.13 Ušće Shopping center. Photographer Biljana Arandelovic. Rights reserved

Fig. 3.14 Ušće Tower 2 construction site. Photographer Biljana Arandelovic. Rights reserved

about 22,000 m^2 will be office space for rent), is planned as the twin tower of Ušće Tower, named Ušće 2. It is designed as an office building according to energy-saving standards, with a restaurant on the top floor and a cafe and bank on the ground floor. This is one of the most attractive locations in Belgrade, having in mind the future development plans for New Belgrade as a business district. The conceptual design of Ušće Tower 2 (Fig. 3.14) was made by Chapman Taylor studio, while the project for the building permit was developed by local Belgrade firm VMS (MPC Properties).

3.5 The Museum of Contemporary Art

The Museum of Contemporary Art is the most important cultural facility in New Belgrade, and it is also an important architectural masterpiece by famous Yugoslav architects Ivana Antić and Ivanka Raspopović. The museum building was built in the Ušće neighborhood between 1961 and 1965 after it won a 1960 competition. The idea of establishing a Modern Gallery in Belgrade, which in the final stages of its implementation was renamed the Museum of Contemporary Art, was not completely new.

The initiative for it came in 1951 (though the idea was much older), when this art institution was founded by modernist painter, art critic and historian Miodrag B. Protić. His main objective was to collect twentieth-century Yugoslav and Serbian art and to find an outstanding location for its promotion (Popadić 2009).

The Museum of Contemporary Art (Figs. 3.15 and 3.16) has the largest collection of Yugoslav art from the twentieth century. Protić should be credited for this distinguished location, at the confluence of the two rivers. The building's location, volume, and modern architecture are striking when seen from Belgrade Fortress, on the other side of the river. The building is located in the central zone of New Belgrade in an area of parkland, and it blends perfectly into the environment. The building was one of the first purpose-designed museum buildings in the former Yugoslavia. It is also one of the first buildings planned in New Belgrade, besides the Central Committee building of the Communist Party of Yugoslavia (CK KPJ), the Presidential Government building, and the high-standard Hotel Yugoslavia. The first competition for the Modern Gallery on the left bank of the Sava River was announced in 1948 on the basis of Dobrović's urban planning concept of New Belgrade (more about Dobrović's urban plan in the chapter: Belgrade's urban plans through history). The museum building is located opposite the Sava amphitheater in the old city where, on the basis of a competition in 1948, the location for the Yugoslav Opera was planned.

A special feature of the design is its successful connection between the exterior and interior, with a big square for events. The space in front of the building is harmoniously designed, and a sculpture park in front of the building is an important part of this urban setting. The cubic volume of the monumental building is merged into an integral structure, without redundant detail, like a polymorphic crystal. The original design solution proposal was different and influenced by New Brutalism, a British movement. The first proposal made was a series of sharp-cornered crystalline volumes in exposed concrete and brick, but after a suggestion by Protić it was changed, and polished marble was used instead. This change gave a more impressive look.

It is interesting to mention the American influence on the development of the Museum of Contemporary Art in New Belgrade. Miodrag B. Protić, its creator, was invited by the Ford Foundation to spend a couple of months in New York in 1962, while the museum was under construction. During his stay, he studied the organization of New York's great museums, especially the Museum of Modern Art, and

Fig. 3.15 The Museum of Contemporary Art. Photographer Biljana Arandelovic. Rights reserved

Fig. 3.16 The Museum of Contemporary Art. Photographer Biljana Arandelovic. Rights reserved

upon his return he applied the knowledge he gained in New York (Kulić 2014). The concept of having a permanent exhibition setting in Belgrade represented, for the modern art of Yugoslavia, what MoMA's permanent exhibition was for international modernism: "an authoritative, canonical view" (Kulić 2014). The museum thus enabled the Yugoslav audience to have a broad overview of international modernism, allowing them to keep in touch with cultural developments and trends in the world. In this way, Yugoslavia was put side by side with most developed countries.

The history of the building was divided into three stages: the first, from 1965 to 1993, was a period of stability, the second from 1993 to 2000, a period of degradation, and the third, initiated in 2000, a period of consolidation, revitalization, and reconstitution. Later events can identify the end of the third period in 2017. The Museum of Contemporary Art was closed in 2007, when the first phase of reconstruction and renovation started. A new stage in the life of the Museum of Contemporary Art has started very ambitiously, and preparation is underway for its 2019 flagship event (Popadić 2009).

A retrospective Exhibition by Belgrade-born internationally acclaimed performance artist Marina Abramović will be organized at the Museum of Contemporary Art from September 2019 to January 2020, marking her return to Serbia, where she had her last solo exhibition in 1975. The retrospective exhibition *Cleaner*, which will take place in seven cities in Europe (Stockholm, Copenhagen, Oslo, Bonn, Florence, and the Polish city of Torun), will end at the Museum of Contemporary Art in Belgrade. *Cleaner* is a project of international importance, which will symbolically mark the return of the Museum of Contemporary Art to the world art scene, after being closed to the public for 10 years and reopening in October 2017 after a long process of reconstruction.

Since the reopening, there has been a constant stream of visitors to the museum. The reason is simple. Belgrade is reestablishing itself on the world stage after years of crises. Belgrade is ready to be a Southeast European metropolis of the twenty-first century, and it has a unique cultural scene to show the world. Also, the aim is to bring the world to Belgrade. Culture and the creative industries are logical steps.

3.6 Hotel Yugoslavia

Hotel Yugoslavia (Figs. 3.17, 3.18, and 3.19) was one of the first three significant facilities planned in New Belgrade, next to the Central Committee building and the Presidential Government building. The hotel was originally supposed to be called Hotel Belgrade and was imagined as a hotel that would represent the state. The competition for the hotel was announced in 1947, and the first prize went to a Croatian team of modernist architects: Lavoslav Horvat, Mladen Kauzlarić, and Kazimir Ostrogović. Their project proposed an uncompromisingly modernist design for the building. The proposal was an appropriate idea for New Belgrade as a contemporary city. The location for the modern hotel—on the Danube Promenade in Block 11, positioned next to the Zemun railway station—was determined in the Preliminary Plan of New Belgrade from 1948 by Nikola Dobrović.

Fig. 3.17 Hotel Yugoslavia in 2010. Photographer Biljana Arandelovic. Rights reserved

Construction of the hotel began in 1948 but was already suspended in 1949, due to the economic crisis resulting from the political crisis during the break with the Soviet Union and Warsaw Pact countries. The works continued in 1960, but according to a revised project by one of the architects from the previous winning team, Lavoslav Horvat. The construction lasted until 1967 and in the following year, the hotel was finally opened to visitors. The interior was completed according to projects by leading Belgrade architects Ivan Antić, Vladeta Maksimović, Mirko Jovanović, and Milorad Pantović (Blagojević 2007).

At the time of opening, it was one of the most exclusive locations in Belgrade and the largest hotel in the whole of Yugoslavia (with seven floors, over 600 rooms and suites, a large restaurant with 600 seats and a smaller restaurant with 200 seats as well as other supporting facilities). The iconic Hotel Yugoslavia was considered to be one of the finest implementations of the ideals of modern architecture in New Belgrade. It has witnessed different events that have shaped Yugoslavia and New Belgrade, and has had its own ups and downs, from being an elite place during the 1970s when it hosted world leaders such as two American Presidents, Richard Nixon and Jimmy Carter, and Queen Elizabeth II, until its fall in the 1990s, when it was hit and partially damaged during the NATO bombing in 1999.

The hotel gradually decayed during the 1980s and 1990s, which led to its final closure in 2000, and one decade later to its privatization. Once grandiose, this object has suffered a similar destiny to the country after which it was named. The

Fig. 3.18 Hotel Yugoslavia in 2010. Photographer Biljana Arandelovic. Rights reserved

current development of Hotel Yugoslavia is connected to the Kempinski hotel chain, which purchased the site of the hotel in 2010.

The hotel is part of a larger high-end mixed-use development project on a surface of 190,000 m^2, the Riverside project. The project area includes the right bank of the Danube River and one of the most popular promenades in Belgrade,

Fig. 3.19 Hotel Yugoslavia in 2018. Photographer Biljana Arandelovic. Rights reserved

located on the edge of a 140 ha natural park. The hotel will be the highlight of the project, but at present it is still awaiting its final reconstruction. Additionally, two new towers will be constructed that will host residential apartments and offices.

3.7 Sava Convention Center

The Sava Convention Center is an international congress, cultural and business center, and one of the first postmodernist buildings in Belgrade, designed by architect Stojan Maksimović. It was built at a time when the international style was dominant in New Belgrade's development. It is located in New Belgrade Block 19 and is defined by Vladimira Popovića Street to the east, Milentija Popovića to the west, and Arsenija Čarnijevića Boulevard to the south.

During the 30 years of New Belgrade's construction, most of the buildings belonged to the international style; however, the basic concept of the Sava Center and the Intercontinental Hotel (part of the congress center) was postmodernism. Sava Center (Fig. 3.18) was part of a larger vision and was planned to be part of the Friendship Center project—related to the active role of Yugoslavia in forming the Non-Aligned Movement and promoting itself worldwide on the international stage.

Yugoslavia, with Tito as its leader, had great ambitions on the international scene. One of the diplomatic highlights of Tito's foreign policy was his success in bringing the Conference on Security and Cooperation (CSCE) in Europe to Belgrade in 1977. This event was important as a forerunner to the Organisation for Security and Cooperation in Europe (OSCE). Belgrade had no convenient modern congress center that could accommodate important international conferences, but since the need for such a facility was immense, the Government took a foreign loan to fund the construction of Sava Congress Center.

Sava Center's construction began in 1976 and Tito opened the big hall on May 15, 1977, although not all the works were completed for this opening. Construction took just 1 year, before hosting the first preliminary meetings of CSCE in June 1977. When construction was completed, Sava Center was a unique facility in this part of Europe, both in its versatile capabilities and its modern architecture. Before the conference, a team of designers led by architect Stojan Maksimović (Manević 2006) was sent abroad in order to tour world-famous examples of this type of architecture, because it was relatively new.

The facility was named the Sava Center after its location, which is adjacent to the Sava River. This peculiar concrete structure with a high-tech glass overlay was designed under the influence of early modernism, following the example of James Stirling and Kevin Roche. The reflective hi-tech glass facade coating followed the latest architectural trends. Thanks to the absence of classical pillars and walls it was possible to get away from the classical form of the space and reach the unbounded interior. In this way, the transparency of the structural elements and installations was also achieved. The architect turned the peripheral areas into a garden and incorporated decorative vegetation with water elements that pass through the space.

> With its exposed trusses and colourful ducts and pipes, the building followed closely in the footsteps of the Pompidou Center in Paris, which opened only one month before the Sava Center. With its sloping glass walls, it also demonstrated a passing resemblance to Copenhagen's recent Bella Center (later the site of the 2009 UN Climate Change Conference), which caused Belgrade's architectural grapevine to whisper of plagiarism. (Kulić 2014: 144)

On the other hand, Sava Center had peculiar organization and an original interpretation of international tendencies. This was a great chance to show the capacity of the domestic construction industry, which would once again confirm the country's success and point to the fact that its place is on the world stage.

Sava Convention Center's construction had to be versatile, due to the lack of concert halls, cinemas, and commercial space in Belgrade. For this reason, the building task was particularly difficult for Stojan Maksimović, the Sava Center architect. The whole complex is a unique urban and architectural entity that extends over 120,000 m^2. Its construction was a three-stage process. In the first phase, which lasted from 1976 to 1977, the basic congress functions were completed. Only 11 months were needed for building this major congress facility. The final result was fantastic, and event spaces today serve the same purpose as when the structure was completed. In second phase, which was during 1978, a large hall with 3,700 places was built (Arsić et al. 1997; Mrenović 2016).

Finally, in the last phase in 1979, the high-class hotel Beograd InterContinental, with 500 units, was built by the same architect, Stojan Maksimović. This luxury hotel was operated by InterContinental, a subsidiary of Pan American World Airways. The management agreement was signed in 1992 (Arsić et al. 1997). Functionally, the hotel is part of Sava Convention Center. The hotel was originally designed as a round tower but the investors proposed a standard parallelepiped volume. Hotel Beograd InterContinental changed its name with its change of ownership in 2012 into Belgrade Crowne Plaza. The hotel functions as a completely separate entity and occupies an area of 36,000 m^2. Hotel Hyatt (Figs. 3.20 and 3.21) joined the neighborhood in 1990, as part of the international hotel group Hyatt Regency, located around two hundred meters from the InterContinental hotel.

Ušće Tower, the former Central Committee building (CK), is also across the street. The whole area is symbolic. Some events which were held in the Sava Center refer to Belgrade's connection with the "First World" and Western financial philosophy. One such event was the 1979 annual meeting of the World Bank and the International Monetary Fund. On the other hand, Yugoslavia was active as one of the leaders of the Non-Aligned Movement of countries. Hotel Beograd InterContinental is a continuation of the same architectural ideas that come from the convention center. The synthesis of inner and outer space is clearly visible.

Sava Convention Center is an independent unity which, besides conference rooms, has restaurants, coffee shops, shops, and also all kinds of different services such as an emergency clinic, a hairdresser, sauna, bank, post office, information center, and agency services.

Fig. 3.20 Hotel Hyatt at Milentija Popovića Street in New Belgrade in 2019. Photographer Biljana Arandelovic. Rights reserved

Fig. 3.21 Hotel Hyatt at Milentija Popovića Street in New Belgrade in 2010. Photographer Biljana Arandelovic. Rights reserved

Architect Stojan Maksimović created the postmodern possibility of partitioning the large hall with panels aimed at creating multiple smaller halls (Mitrović 2012). The highlight of this construction is the fact that Sava Convention Center was nominated for the Pritzker Prize for architecture in 1979, the same year in which this prize was established.

One of the biggest successes of this ambitious project is the longevity and maximum utilization through its four decades of existence. Sava Convention Center has hosted several high-profile political and economic events since its opening on May 15, 1977, among others the 9th Summit of the Non-Aligned Countries held in 1989. Furthermore, this is a prime location for all kinds of performances including film festivals, concerts, and other events.

On December 15, 2017, the City of Belgrade announced a new tender for a public–private partnership for the reconstruction and management of Sava Convention Center (Figs. 3.22, 3.23, and 3.24). The city government is looking for a partner for a joint venture who will lead the reconstruction and manage the center. The city of Belgrade would own 51% of the capital, and the private partner 49%. This process is still ongoing. The complete renovation of the congress infrastructure is planned, with the possibility of building new facilities. After years of crisis, the building requires renovation by contemporary standards (this refers primarily to energy efficiency).

3.8 The Construction of New Belgrade's Central Zone Blocks

New Belgrade's blocks symbolize the policy of social welfare at the time of Yugoslavia and the time of massive industrialization in the construction industry. Each one of New Belgrade's block was constructed with, besides the primary function of housing, all the necessary supporting facilities: schools and kindergartens, as well as supply facilities. The construction of New Belgrade can be identified in three phases. The period between the competition and its realization was accompanied by social and political changes, from creating a one-party system through to the transition to self-governing social ownership, according to the 1963 Constitution. The six blocks of the Central Zone were built in the 1980s. The central blocks in the central area, Blocks 24, 25, and 26, were not built according to the original idea and they were built at the end of the 1980s (some even later, by the breakdown of Yugoslavia). The aesthetics of the blocks have been disputed many times, and they have been emphasized as inhuman cold gray urban megastructures. New Belgrade is certainly a unique territory, a new city within a city that marks an epoch of post-war urban development. New Belgrade's post-war modernism has, above all, the features of International style (Djordjević 1960; Ristanović 2009).

The first phase began with the construction of Tošin Bunar (1947–1950) and the Student City (1948–1950) settlements. Tošin Bunar is one of the first in a series of

Fig. 3.22 Sava Convention Center. Photographer Biljana Arandelovic. Rights reserved

Fig. 3.23 Sava Convention Center. Photographer Biljana Arandelovic. Rights reserved

Fig. 3.24 Sava Convention Center. Photographer Biljana Arandelovic. Rights reserved

mass constructions. The state wanted to resolve the housing crisis through mass construction in the new city area. The buildings were made of bricks, without central heating, using several standard designs. The architectural and urban projects were made by Ljubo Ilić and Vido Vrbanić.

The second phase began in 1959 and lasted until 1963 with the construction of Block 1 and Block 2 designed by Branko Petričić, which were so-called experimental settlements. During this period, the building of Fontana Community Center was created in Block 1. Block 1 was the first block to be designed and built based on the Master Plan of New Belgrade 1958. This block symbolically became number one when the idea to number the blocks was first used during the preparation of the Regulatory Plan of New Belgrade. The blocks follow Le Corbusier's ideal dimensions of 400 × 400 m, and the buildings are arranged to give balance, access to sunlight and open vistas, and they are grouped so that in the center of a block there is an area for recreation, sports courts, and a playground (Gligorijević and Graovac 2018).

The construction of Blocks 1 and 2 and the Fontana complex announced a new building concept, industrialized construction, along with a construction system by Branko Žeželj (IMS Žeželj system). The IMS system consists of basic construction elements, columns, and slabs with prestressed cables within them, often used in a module of 4.2 × 4.2 m. It was designed to be more flexible than any other kind of prefabrication. This system was used for much of the building process for the blocks in New Belgrade, and it became one of its construction symbols (Stevanović

2015). The main asset of the IMS Žeželj system was its great flexibility, with a system for both the interior and exterior designs.

The third period began after a 1958–1959 competition to find a solution for the first residential block in the Central Zone. This was the beginning of a long process. When the competition was over, the Council of Urban Planning set new rules, according to which drafting of the plan for this area was entrusted to the Belgrade Planning Institute, which formed a working group of architects and urban planners to do the job. The plan set the main contours of the central zone of New Belgrade, which still exist today. Based on this decision, Milutin Glavički and Uros Martinović created a Regulatory plan in 1962 that served as the basis for the development of detailed plans and zoning requirements for all future blocks and individual buildings in New Belgrade (Blagojević 2007; Glavički 1965).

Block 21 was the first housing project implemented in the period 1962–1966 in New Belgrade (close to SIV Palace). Five more new blocks followed the construction of Block 1, all before 1979, forming the central zone of New Belgrade: Block 22 and Block 23 (1976), Block 28 (1971), Block 29 (1973), and Block 30 (1979). However, the central zone of New Belgrade was just partially implemented. It is important to underline that from the point of view of urban planning, the central zone of New Belgrade was on the same level as the old core of Belgrade. The idea of planners was that this stretch should not become a satellite housing area but rather be equal to Belgrade's old center (Vesković and Jovanović 2018).

Over time, the plan to establish a monumental zone in New Belgrade was dropped, which accelerated the construction of the highway through New Belgrade and development of Block 24 in the period 1984–1989 (according to the project by Bogdan and Vlada Slavica).

Fontana Community Center (Fig. 3.25), the administrative unit that forms a square at Pariske Komune Street in Block 1, is an old symbol of New Belgrade, built in 1963 by Uroš Martinović. The complex is listed as a cultural monument representing modern Serbian architecture, and it was recently reconstructed (2010–2013). Fontana, a multifunctional facility, is a successful attempt at the urban and architectural humanization of New Belgrade's uniform blocks, making them more suitable for human needs. Urban motifs such as passages, porches, columns, canopies, a water fountain, and an atrium break the monotony of the orthogonal urban networks of the residential part of New Belgrade. The popular cinema in Fontana was closed in 1991 and reopened in 2012 after renovation, together with other new service facilities.

Block 21 is particularly significant as it was the first block completed, and it is considered as the model for the central zone of New Belgrade. This block was designed with the goal of establishing a suitable environment for the most important structures, the objects of the state and the political administration. Initially, Block 21 was built for the housing needs of the Yugoslav People's Army.

Blocks 22 and 23 were the subject of a competition from 1968, constructed according to the winning solutions of Božidar Janković, Branislav Karadžić, and Aleksandar Stjepanović, with abundant use of concrete. The architects were fully dedicated to forming typical block cells, which also incorporated a striking visual

Fig. 3.25 Fontana Community Center today. Photographer Biljana Arandelovic. Rights reserved

Fig. 3.26 Block 22. Photographer Biljana Arandelovic. Rights reserved

composition and can be seen from distant parts of the city, as well as the opposite bank of the Sava (Arsić et al. 1997). In addition to the use of international style, brutalism can also be identified as a style visibly present in New Belgrade's blocks, especially in Blocks 22 and 23, where concrete is the main decorative element on the facades (Fig. 3.26).

Blocks 25 and 26 were the subject of a competition in 1998, won by Svetislav Ličina. This winning project was the basis for the Regulatory Plan for Blocks 25 and 26. The next competition for these two blocks was in 2007, with a detailed regulation plan being adopted in 2016.

3.9 New Belgrade Blocks 61 and 62

New Belgrade Blocks 61 and 62 were built by architect couple Milenija and Darko Marušić (with Milan Miodragović) between 1972 and 1976, after their design solution won a 1968 competition (Marušić 1972). Blocks 61 and 62 (Fig. 3.27) are based on a detailed urban plan adopted in 1965. The Marušić couple were important players in developing residential architecture in Yugoslavia, with special emphasis on Belgrade's architecture in the 1970s and 1980s. They are the architects of

Fig. 3.27 Residential Blocks 61 and 62 New Belgrade. Photographer Biljana Arandelovic. Rights reserved

famous Belgrade settlement Cerak Vinogradi (Cerak I and Cerak II), the first fully modern residential area, placed under the authority of the Institute for the Protection of Cultural Monuments in January 2019. The Cerak Vinogradi settlement has not changed its original appearance from the time when it was built in 1987 (construction lasted for a full decade). The Cerak Vinogradi residential settlement was part of the exhibition *Toward a Concrete Utopia: Architecture in Yugoslavia, 1948–1980* held in the MoMA Gallery in New York in 2018 (Stierli and Kulić 2018).

The Marušić couple belonged to the Belgrade School of housing, where the idea of the "Belgrade apartment" dominated. The Belgrade School tried to harmonize "the ideas of flexible, convivial and individual space" (Stevanović 2015). New Belgrade Blocks 61 and 62 were a perfect example of where this approach was implemented. In order to avoid monotony, the height of buildings varied from 4 to 20 floors. There are two object types, type A and type B. Type A can be found in groups of three, where each contains six lamellas carrying staircases and four apartments around them on each floor. Type B buildings come in pairs, each with four lamellas. The IMS Žeželj system, with standard dimensions of 4.2 × 4.2 m (and also 3.6 × 4.2 m), was used here as well. Prefabricated concrete panels were also used for ceilings, interior walls, and facades.

Blocks 61 and 62 represented "a shift in functional design, and also one of the clearer examples of organizing two-wing buildings in Serbian multifamily housing." The Marušić couple "expressed an incompact concept of grouping and positioning the blocks," they were "thinking about the buildings as smaller neighbourhood units, thus building an ambience with a recognizable identity" (Milašinović-Marić and Vukotić-Lazar 2017).

What the architects wanted here was to propose a distinctive morphology for New Belgrade's residential blocks. They enhanced the organization of the apartments to emphasize the flow-circular scheme, a recognizable element of the "Belgrade apartment."

3.10 New Belgrade Block 12

Block 12 is one of the central parts of New Belgrade and it is located in the area between the Palace of Serbia, Hotel Yugoslavia and Blocks 30, 31, and 11. Also nearby is a green belt stretching along the Danube. Traffic is well connected with the rest of the city through a large number of public transportation lines, and two wide boulevards (Nikola Tesla and Mihajlo Pupin Boulevards) (Fig. 3.28) which are among the busiest routes in Belgrade. The immediate vicinity of the highway is another advantage, making the location of Block 12 with regard to traffic one of the most favorable positions in Belgrade.

Block 12 is one of the newer blocks in New Belgrade, built in the era of post-socialist transition under the leadership of two architects, Milutin Gec and Mario Jobst, who worked within the Energoprojekt Company. It consists of a residential and business complex with nine buildings and common open space between the buildings, located at Mihajlo Pupin Boulevard, 10. The buildings cover

Fig. 3.28 Mihajlo Pupin Boulevard. Photographer Biljana Arandelovic. Rights reserved

an area of 180,000 m^2 and are marked by letters (A, B, C, G, D, E, W, Z, and I). Construction began in 1991, with the first four buildings being completed in 1994, and the entire project completed in 2002 (Energoprojekt 2011).

Because of Block 12's commercial functions, it is popularly referred to as the YU business center, abbreviated YUBIZ (from Yugoslavia), since it was designed at a time when Yugoslavia existed and the nickname remained.

The Old Telephone Exchange building by famous Serbian architect Branko Tanazević, built in 1908, is among the first examples of Serbian architecture which received a more modern treatment of its Serbian-Byzantine style and Moravian school ornamentation. It was a direct inspiration for architect Mario Jobst in his design of the YU business center complex (Figs. 3.29 and 3.30). The architecture of Gaudi is cited as another source of inspiration for Jobst.

The steel construction on the facade forms the shape of sails, aiming to simulate the connection with the nearby Danube River. The architect used a striking shade of red on the facade of the building, which is characteristic of his work, and this later became popular on the Serbian architectural scene (Jobst 2000). The use of red is also linked to the works of Tanazić. Jobst sought to bring dynamics into his architecture and to swing static objects into motion through the introduction of certain sculptural forms. The color red further emphasizes this (Alfirević 2012).

Fig. 3.29 YU business center complex by Milutin Gec and Mario Jobst. Photographer Biljana Arandelovic. Rights reserved

Fig. 3.30 YU business center complex by Milutin Gec and Mario Jobst. Photographer Biljana Arandelovic. Rights reserved

3.11 Philharmonic Building Development Project in Block 13

The winners of the competition for the design of one part of Block 13 (held in 2016), which envisages the construction of a new concert hall of the Belgrade Philharmonic, were architects Dragan Marčetić and Milan Maksimović (Figs. 3.31, 3.32, 3.33, 3.34, 3.35, and 3.36). This is one of the projects from the new vision of New Belgrade, which at the same time can be considered as part of the urban development of the coastal areas of the Sava River, thanks to its close vicinity to the river.

Fig. 3.31 Belgrade Philharmonic project. Image provided by Dragan Marčetić and Milan Maksimović. Rights reserved

Fig. 3.32 Belgrade Philharmonic project. Image provided by Dragan Marčetić and Milan Maksimović. Rights reserved

The construction of the Belgrade Philharmonic is associated with the development of Block 13. Within this block is also an important historical building, the SIV Palace. The conceptual approach for the Philharmonic building complex stemmed from a desire to preserve the existing character of the natural environment, which is directly linked to the waterfront on the Sava riverside. The structure of glass and steel looks like it slowly emerges from the ground, and this impression is further underlined by its green roof.

Greenery is an essential element of the new cascading form of the Philharmonic building, and it is intensified on the northern side of the new urban composition with the aim of creating a green oasis. The roof is harmonized with the natural

Fig. 3.33 Belgrade Philharmonic project. Image provided by Dragan Marčetić and Milan Maksimović. Rights reserved

Fig. 3.34 Belgrade Philharmonic project. Image provided by Dragan Marčetić and Milan Maksimović. Rights reserved

environment of the waterfront. There are new areas within the complex that were designed to be active areas of greenery and recreation. The commercial shopping mall is directly connected with the Philharmonic building. According to this green design concept, the building is designed with natural materials. The facade cladding and paving is of natural stone, while natural wood is used for the interior halls. The project has not yet been implemented (DAB 2018).

Fig. 3.35 Belgrade Philharmonic project. Image provided by Dragan Marčetić and Milan Maksimović. Rights reserved

Fig. 3.36 Belgrade Philharmonic project. Image provided by Dragan Marčetić and Milan Maksimović. Rights reserved

3.12 New Transport Hub in Block 42 of New Belgrade

One of the strategic development plans for Belgrade is the relocation of the bus and train stations from the old center in Belgrade to New Belgrade Block 42. The plan for the detailed regulation of the bus and train station in Block 42 was adopted in 2016. The location was investigated by means of two urban architectural competitions, in 1995 and 2014. The aim of the competitions was to select the best urban and architectural design for the construction and development of the new stations, but also to resolve the issue of developing a complex transport network and to regulate all types of traffic issues (public city transportation, vehicle, and bicycle traffic).

The winning solution by Milan and Vladimir Lojanica was the basis for developing a plan. Block 42 was strategically chosen for the relocation since it is the crossing point of the two axes of New Belgrade. The first of these is the historical axis from the Federal Executive Council building toward the Sava, and the second is the new development axis along the railway and road corridor. Two aspects were essential for designing the identity of the location: the feasibility of constructing tall buildings and creating a representative public space in accordance with New Belgrade's urban structure. Within the context of the conditions in the location and the most efficient use of city building land, the design includes a business and commercial complex. The whole project has received significant public investment for the strategic development of New Belgrade.

According to the design by Milan and Vladimir Lojanica, the new main bus station in New Belgrade will have 15 incoming and 58 outgoing terminals, while the future main railway station will have 8 platforms. The project is one of the biggest city hubs, covering a total area of 21 ha, which will be a center for public transport logistics. When fully completed, the whole complex will be built on a total of 180,000 m^2 (including the station buildings, office, and commercial space and three towers of 23 floors). The design includes clearly divided entities, which will all be well connected in terms of communication. According to the detailed regulation plan, the height limit was up to 150 m but the winning design proposed not to go up to a maximum of 150 m, but instead to distribute the square footage across three buildings of not more than 100 m in height. The towers are positioned at the edges of the block to delimit the space and provide an elegant accent of height.

The central focus of this complex will be the new transport terminal with the bus and railway station, and the main facade of the station building will be designed in relation to a large access plateau for pedestrians. A harmonious urban ambient will be formed by the following elements: the towers as visual highlights, the station building and a wide plateau for pedestrians, and wide areas of greenery. The main station building will have an expressive canopy, which further emphasizes the building's sharp horizontal axis. This is a good opportunity to raise the standard of the aesthetics of New Belgrade. The complex is currently in the development phase. On the basis of the design by Milan and Vladimir Lojanica, a detailed regulation plan will be made, after which the development will continue with the first project phase. The bus station is the most important part of the project and will be implemented in the first phase (DAB 2018).

3.13 Block 18 Development Project

An urban and architectural design competition for Block 18 in New Belgrade was held in 2016. The theme of the competition was extremely complex, given the requirements for the conceptual and program framework of the new city structure at the key site of the Sava amphitheater. The location is an extension of the central axis of Zoran Djindjić Boulevard and the Sava Center and includes the green belt near the left bank of the Sava, which now stretches from Ušće to Block 18, and in the future will go as far as the western part of New Belgrade. This block is specific because behind it are some of the most successful architectural blocks in New Belgrade: 21, 22, 23, and 19a, from different periods and with different urban matrices. On the right bank of the Sava River is the ongoing development of the Belgrade Waterfront project, with a spatial structure which, when it is completed, will dominate the whole area of the Sava coastal area. In terms of the landscape, Block 18 is a counterpoint to the planned Belgrade Waterfront project.

In spatial and functional terms, the location is important both as an excellent location for housing and as a link between crucial urban routes. The solution requires a balance between urban, public, and private housing space of high density. The first prize in the competition went to a team of authors: Vanja Panić, Marko Vesković, and Aleksandar Knežević, for whom one of the main objectives was to emphasize pedestrian zones. The project area covers 46.8 ha. The already existing zoning scheme is going to be transformed from free, semi-atrium, and line blocks, into closed traditional blocks, which form the front of the shoreline. Towers are anticipated in the vicinity of the highway, with additional two towers along Vladimira Popovića Boulevard that will frame the urban zone of Block 18.

The main attraction of this area will be Opera Square, where the future opera building will be situated. Further connection with the coast in the old town across the river will be made by means of the promenade and pedestrian bridge. Within the plan, facilities are envisaged for housing, education (school and student campus), art and cultural events, health, trade, business, and commercial activities. Business suites will be on the outer edges of the block while a hotel will be on the coast. An institute for art and culture, the opera building, a museum, and a gallery are also projected. Commercial and shopping facilities, restaurants, and cultural facilities are included in the pavilions on the promenade. Envisaged around the block and along the main roads are mainly commercial facilities, while the central part of the block will be predominantly occupied by housing. However, housing zones are situated in all parts of the block with the commercial facilities.

The planned Block 18 reconstruction gives a modern context to New Belgrade's development and improves the Sava coastal zone. The concept of a traditional urban structure with street edge blocks, squares, and public greenery is adapted to modern times. The major axis is highlighted by placing the street in the center of the block perpendicular to the river and extending this direction over the pedestrian bridge. This clearly enables direct connection with New Belgrade on one side and Belgrade Waterfront project on the other side of the Sava River. The two banks are connected by a functional scheme of pedestrian and bike trails that stretch across the pedestrian bridge over the river.

The height of the buildings is suited to the street profiles and city structure as a whole, with the accent on the corners of the locations along main roads. Medium-high blocks (up to 30 m), which are lower than those of the hinterland, descend toward the coast and create a counterweight toward the altitude accent of the Belgrade Waterfront buildings on the opposite bank of the Sava (DAB 2018).

3.14 New Belgrade Perspectives

New Belgrade is the most famous post-socialist city in this region. Many ambitious goals that were set during the socialist era have never been achieved, and New Belgrade has entered a new post-socialist and post-Yugoslavian era without a clear vision of development, but on the other hand it has been developed in line with the new circumstances in Serbia. In a country in transition, nothing can be accurately predicted. The changes that have occurred in post-socialist Belgrade cannot be classified as planned, but rather they have arisen as a chaotic reaction to the unexpected transformation of Serbian society. Since the 1990s, New Belgrade has come under the influence of neoliberal urban transformation, with many new construction sites appearing that are not part of the regular planning.

Illegal construction in New Belgrade has become the logical consequence of the events that took place in the city at the beginning of the 1990s. One of was the search for ways to gain more space for use by new private businesses, mainly in the commerce sector. This has been especially noticeable in New Belgrade because of its enhanced residential character. Free spaces on the ground floor of residential buildings proved to be favorable places for the expansion of new business contents. Anarchy in urban planning has directly enabled the construction of new buildings in green areas, which were previously used for recreation. The free market economy has increased the value of apartments in attractive locations, and some owners have decided to rent them for business purposes. Commerce and the service and financial sectors are now firmly positioned in the urban fabric of New Belgrade. Thus, the area of the city that was previously characterized by residential facilities has now become an area filled with commercial and business facilities (Fig. 3.38). After democratic changes in 2000 and the opening of Serbia to the global market, New Belgrade became fruitful ground for new shopping centers, banks, hypermarkets, and branches of financial and insurance companies (Jocić 2019). An important business area was developed between Arsenije Čarnojević Boulevard and Mihajlo Pupin Boulevard which spreads to the west, in the direction of Zemun. The old center of Belgrade has lost its leading role as a business center (Ratkaj 2009).

New Belgrade now has the highest number of employees and obtains more investments than any other municipality in the city. Additionally, it brings about the biggest portion of Belgrade's GDP. The new state of the economy has changed the urban perception of New Belgrade, resulting in a place with a former peripheral position gaining a central place within the city (Jocić 2019).

Its favorable location for the development of new businesses is also supported by data on the rise of retail centers in New Belgrade. The number of retail jobs in New Belgrade has increased four times, from 5,738 in 2002 to 23,051 in 2012. Shopping centers such as Delta City, Ušće, and Mercator have become some of the distinguishing features of contemporary New Belgrade. Furthermore, the total number of people employed in New Belgrade doubled in the period from 2002 to 2011, from 52,600 to 103,000. In the same period, the number of jobs in Stari Grad (the old city center municipality) dropped from 104,000 to 55,000 (Jovanović and Ratkaj 2014).

Another example of a new business development is Airport City Belgrade, which is a vast complex of modern glass and steel buildings built in Block 65 as the first multi-use commercial business park in Serbia (Hirt 2009). The construction of Airport City Belgrade started in 2005 and went on until 2016, when all nine buildings were finished. There is a plan for further enlargement, but no action has yet been taken on this. The complex is spread over 14 ha and is part of the emerging Central Business District. Airport City Belgrade is located near the Euro-Motorway E-75 and in the vicinity of Belgrade airport. In its vicinity is also Belville, a modern neighborhood, as well as the Delta City shopping mall.

The rise of the IT sector in Serbia is another topic directly connected to New Belgrade, because most companies in the industry are based in its central area, in Block 12. Opening Serbia to the world market after the political changes in 2000 has resulted in the development of the IT sector. Location and infrastructure advantages have resulted in the IT industry making its base in Block 12. One of the most successful Serbian IT start-ups, Nordeus, which was established in 2010, is among the companies located in this part of town. In addition, the Microsoft Development Center has been located in Block 12 since 2012. In 2010, 9 companies were active in Block 12, while in 2016 that number rose to 23 relevant IT companies (Jocić 2019).

New Belgrade has still maintained a strong residential function but also become an emerging business municipality (Fig. 3.38). Both functions are very important for its further development. It is necessary to mention another important characteristic, which is that of cultural policy, and it includes the construction of the

Fig. 3.37 New Belgrade. Photographer Biljana Arandelovic. Rights reserved

Philharmonic building and Chinese cultural center. At the location of the former Chinese Embassy in New Belgrade that was damaged during the 1999 NATO bombing, a new Chinese cultural center of 32,300 m^2 is being developed, which is one of the biggest Chinese investments of this kind in the world. The decision to build such a center is not surprising, as China is one of the largest investors in the country. The 32-m-tall building will be a multifunctional facility next to the cultural center, with a business and a residential wing. The building has been under construction since 2017 and is a work in progress.

There is still ongoing controversy surrounding the construction of the opera house. The last competition for the opera house project was held in 1970, and the

Fig. 3.38 New Belgrade. Photographer Biljana Arandelovic. Rights reserved

anticipated location was on the left bank of the Sava in Block 17 (Stojanović et al. 1974). The main polemic was whether the opera house should be situated in the center of the old town or on the territory of New Belgrade. New Belgrade was chosen because of its central position in the future structure of the city. However, the matter of location is still a problem.

New Belgrade has never been fully completed, and it is still a work in progress (Fig. 3.37). It was considered to be utopian project and is a constant topic of polemics as well as an endless inspiration for developers, architects, artists, the government, and investors. It started with enormous ambition more than 70 years ago and is still full of empty spaces and promising locations (Fig. 3.38).

References

Alfirević Đ (2012) Is there expressionism in Serbian architecture? An interview with architect Ognjen Djurovic, Branislav Mitrovic, Mario Jobst. Arhitektura i urbanizam 35:72–83
Arandelovic B et al (2017) Belgrade: imaging the future and creating a European metropolis. Cities 63:1–19
Arsić J et al (1997) Gradjevinarstvo Srbije. Ministarstvo gradjevina Srbije
Blagojević L (2007) Novi Beograd: Osporeni modernizam [New Belgrade, challenged modernism]. Zavod za udžbenike, Beograd
Blagojević Lj (2005) New Belgrade: the capital of no-city's-land. Art-e-fact 4
DAB (2018) Društvo arhitekata Beograda Association of architects Belgrade. http://www.dab.rs/
Djordjević A (1960) Urbanističko rešenje centralnog dela Novog Beograda. Arhitektura Urbanizam 2:3–13
Djurović V (1995) Urbani razvoj 1971–1991, gradska studija za habitat [1971–1991 Urban development, urban studies for habitat]. Centar za urbani razvoj CEP Beograd
Energoprojekt (2011) Energoprojekt – 60 godina uspeha. Monografija povodom 60 godina od osnivanja kompanije. Belgrade (Enegroprojekt)
Glavički M (1965) Regulacioni plan Novog Beograda. Arhitektura urbanizam 32 (Beograd) 5–11
Gligorijević Z, Graovac A (2018) 70 godina Urbanističkog zavoda Beograda. Knjiga 2. Urbanistički zavod Beograda
Hirt S (2009) City profile: Belgrade, Serbia. Cities 26(5):293–303
Jobst M (2000) Stambeno-poslovni objekat G u Bloku 12, Novi Beograd. Arhitektura Energoprojekt 10:13–18
Jocić N (2019) Creative economy in a post-socialist transitional context: spatial distribution of creative activities and their interrelation with urban milieus in city quarters of Belgrade. University of Bamberg Press
Jovanović M, Ratkaj I (2014) Functional metamorphosis of New Belgrade. disP Plan Rev 50 (2014):54–65
Jović J, Djorić V (2010) Traffic and environmental street network modelling: Belgrade case study. Transport 25(2):155–162
Kulić V (2013) National, supranational, international: New Belgrade and the symbolic construction of a socialist capital. Natl Pap 41/1(2013):44
Kulić V (2014) New Belgrade and socialist Yugoslavia's three globalisation. Int J Hist Cult Mod 2 (2):125–153
Kulic V, Mrduljas M, Wolfgang T (2012) Modernism in between: the mediatory architectures of socialist Yugoslavia. Jovis, Berlin
Macura M (1947) Our architecture in the light of the competition for the building of the presidency of FPRY. Arhitektura 3 Zagreb

Maksimović B (1983) Ideje i stvarnost urbanizma Beograda 1830–1941 [Belgrade urbanism ideas and reality]. Zavod za zaštitu spomenika kulture grada Beograda
Manević Z (2006) Stojan Maksimović: Stvaralaštvo. Centar
Marušić D (1972) Blokovi 61 i 62 u Novom Beogradu. Izgradnja special edition 125–133
Milašinović-Marić D, Vukotić-Lazar M (2017) Modernism versus postmodernism as an impetus to creativity in the work of architects Milenija and Darko Marušić. SPATIUM 37:12–21
Mišić B (2011) Palata Saveznog izvršnog veća u Novom Beogradu. Istorija gradjenja. Od Predsedništva vlade FNRJ do Palate Srbija [Palace the Federal Executive Council in New Belgrade. History of construction. From the Presidency of the Yugoslav Government to the Palace of Serbia]. Zavod za zaštitu spomenika kulture grada Beograda
Mitrović M (2012) Arhitektura Beograda 1950–2012. Službeni Glasnik Beograd
MPC Properties. http://www.mpcproperties.rs/uscetowertwo.php
Mrenović L (2016) Arhitektura Sava centra u kontekstu urbanističkog razvoja Novog Beograda [Architecture of Sava Center in the context of urban development of New Belgrade]. ARTUM 3:37–45
Novibeograd.rs Facts about New Belgrade. http://novibeograd.rs/. Accessed 16 Dec 2018
Popadić M (2009) Arhitektura muzeja savremene umetnosti u Beogradu [The architecture of the Museum of Contemporary Art in Belgrade]. Nasledje 10:159–178
Ratkaj I (2009) Prostorno-funkcionalna organizacija Beograda. Geograf. Fak
Ristanović S (2009) Novi Beograd: Graditeljski poduhvat veka [New Belgrade: construction venture of the century]. Kosmos: Beograd
Statistical Office of the Republic of Serbia (2017) Municipalities and regions of the Republic of Serbia
Stevanovíc T (2015) Tools for conviviality: architects and the limits of flexibility for housing design in New Belgrade. In: Lloyd Thomas K et al (eds) Industries of architecture 160–170
Stierli M, Kulić V (2018) Toward a concrete Utopia: architecture in Yugoslavia, 1948–1980. The Museum of Modern Art New York
Stojanović B et al (1974) Urbanizam Beograda 16. Town Planning Institute of Belgrade 6–18
Stojanovic B, Martinovic U (1978) Beograd: 1945–1975: arhitektura: urbanizam. Tehnička knjiga Beograd
Vesković I, Jovanović J (2018) Izgradnja blokova 21, 22, 23 centralne zone Novog Beograda i njihov značaj u okviru kulturnog nasledja Beograda [Building blocks 21, 22, 23 of central zone of New Belgrade and their importance in the cultural heritage of Belgrade]. Nasledje 19:35–51
Vujović S, Petrović M (2007) Belgrade's post-socialist urban evolution: reflections by the actors in the development process. In: Stanilov K (ed) The post-socialist city. Springer

Chapter 4
Belgrade's Waterfront Revitalization

Milena Vukmirovic

Abstract This chapter deals with the waterfront revitalization, presenting built and unbuilt projects that are part of the city's eternal ambition to get down to its rivers, the Danube and Sava. Since the regulation of the Sava and Danube shoreline is one of the priorities for the city's development, this topic is crucial for Belgrade. Unfortunately, Belgrade is one of a few European cities that have not yet managed to successfully integrate their riverfronts into the city core. Until recently, Belgrade's riverfront has been characterized by devastated industrial zones and public promenades that are in poor condition. However, several megaprojects have been initiated in order to change this situation. Defined as priority projects in the City of Belgrade's Development Strategy, they will be presented and discussed in this chapter.

Keywords Waterfront · Megaprojects · Sava · Danube · City of Belgrade

Belgrade was founded on the confluence of two rivers, the Sava and Danube, making them an inseparable part of its image and identity. However, for a long time they were treated as hinterland, barriers, and boundaries. In the wake of new trends in the reconstruction and development of cities at the beginning of the twentieth century, the first thoughts about the need for "lowering the city onto its rivers" and using the waterfronts in an appropriate way appeared. After WWII, the potentials and possibilities of the Sava and Danube rivers became a "versatile preoccupation of the city and all its governmental structures" (Vučelić 1977), which can be seen in the first annual plan for the socioeconomic development of Belgrade. In the 1950 General Urban Plan, the basics for the thoughtful and planned treatment of the Sava and Danube resources were adopted.

The construction of New Belgrade has contributed to the river flows and coastlines receiving urban tissue treatment. This idea was furthered by the construction of Belgrade Fairground and the Danube Port, the transformation of Ada Ciganlija into a modern sports and recreation center, and the displacement of the Sava shipyard and non-hygienic settlements. Everything that was done in this area, from the construction of coastal forts, docks, communal, traffic, and industrial

© Springer Nature Switzerland AG 2020
B. Arandelovic and M. Vukmirovic, *Belgrade*, The Urban Book Series,
https://doi.org/10.1007/978-3-030-35070-3_4

facilities, as well as the definition of further reconstruction possibilities, was aimed at the further urbanization of these parts of the city, and the rational use of river flows and coastal areas. Despite "the return of Belgrade to its rivers," with the exception of work done in the planning and design domain, the coastal area was not preserved, reconstructed or revitalized; on the contrary, it was systematically attacked, burdened, and overburdened with inadequate content and forms.

Like the plans from 1972 and 1985, contemporary strategic and planning documents have inevitably dealt with the theme of rivers and coastal areas. The General Urban Plan 2021 established the strategic framework for the future development of Belgrade coastal areas, recognizing and emphasizing their exceptional importance for the city. The tourist potential of the central city zone stands out, which needs to be activated and complemented with different cultural, sport, and entertainment programs, as well as with the development of city, passenger, and taxi traffic on the river. The strategic orientation of Belgrade is defined by the lowering of the old city center/core to the rivers, which would expand its central urban area.

Belgrade's strategy for tourism development highlights the importance of its coastline for progress in the area of tourism. In defining different tourist products, the strategy states that "the backbone of the tourist product 'city break' should be a major tourist pedestrian zone of Belgrade with the Sava and Danube as aqua boulevards" (Palgo center 2011). The tourist product listed by the strategy combines a variety of programs such as culture, sport, history, ecology, river experience, nightlife, special interests, events, nautical sports or boating, and shopping.

The City of Belgrade development strategy for 2021 (RAREI and Palgo center 2017) defines the implementation of projects and plans related to the rivers and coastal areas as a priority for improving the identity of Belgrade. Accordingly, planning, designing, building, and promoting the basic elements of the identity of the city of Belgrade will be achieved through: the construction of the Belgrade Waterfront project and the arrangement of Sava promenade; the preparation and adoption of the regulation plans for Blocks 18, 18a, 68, and 69 in New Belgrade; the development and adoption of the detailed regulation plan for a sports and recreational center on the Danube riverbank in Grocka; the preparation and adoption of the detailed regulation plan for the area between the heating plant complex, the Sava embankment, Gazela Bridge, and the Sava River in New Belgrade; the development and adoption of the detailed regulation plan for the area of Ada Huja; and the preparation and adoption of the detailed regulation plan for the Old Fairground (Staro sajmište) complex. Based on the above, it can be concluded that the objectives are carefully defined in accordance with the available resources. In the coming period, priority will be given to the construction of Belgrade Waterfront, while other locations will be mainly planned and defined.

In the study of the Belgrade coastal area (2007), for the first time, the area of the Belgrade coastal area was treated as a unique whole from every aspect. The key result of this study was defining the boundary of the wider coastal zone, the zone of influence, and the zone of wider interest (Fig. 4.1). Accordingly, all the research was concentrated on the so-called spatial framework of about 16,950 ha, covering the entire aquatorium, as well as part of the territory whose width varies from

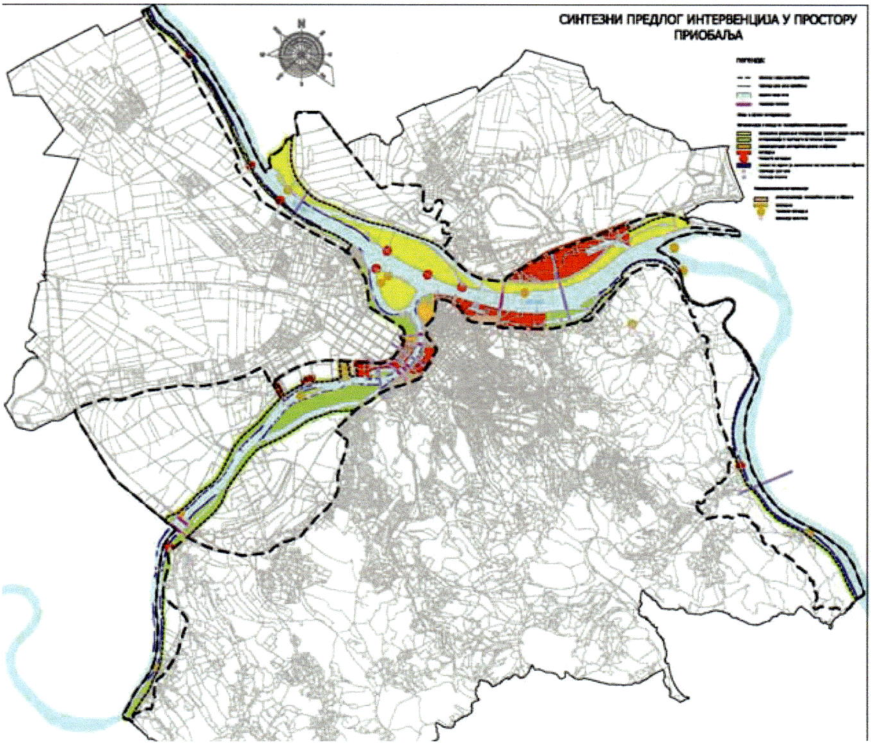

Fig. 4.1 Study of Belgrade coastal area I phase (2007). Archive: Belgrade Urban Planning Institute

several tens of meters to several kilometers in the hinterland. By analyzing this defined space, focusing on a belt that includes the area along the coastline where the activities are most intense, therefore making it the subject of much interest from the aspect of potential and possible interventions, the border of the narrow coastal zone has been formed, which includes an area of around 5,587 ha.

On the other hand, the synthetic conclusion of the study was "that, although the Belgrade coast makes itself an invaluable asset with unbeatable potential for Belgrade, the advantages of its position have never been sufficiently exploited in the entire history of the city's development, nor have they been adequately evaluated." In addition, the general attitude was that the coastline is in an unacceptable condition, with inadequate use of the water and the coast, while the most valuable, central parts are almost completely degraded.

The city gains its identity mainly from the central urban zone and the position of Belgrade at the mouth of the Sava and Danube. But in the future, the emphasis will be on the restoration of the existing urban structures, especially the central parts, and their compactness, identity, mixed uses, green areas, and riverbanks, as well as the facilities and complexes of capital value. The natural and cultural heritage will have a special role in this, and the city will try to draw maximum value from this part of its capital, thus competing in a special manner with other European capitals. Strengthening its competitiveness is seen as a fundamental goal as well. This can be achieved by activating dormant resources, as well as mobilizing local capital, inactive so far, and by other means.

Since regulating the Sava and Danube coastal area is one of the priorities for the city's development, this topic is crucial for Belgrade. Unfortunately, Belgrade is one of the few European cities that have not yet managed to successfully integrate their riverfronts into the city core. Until recently, Belgrade's riverfront has been characterized by devastated industrial zones and public promenades that are in poor condition. However, several megaprojects have the potential to completely change this situation. They are recognized within the framework defined in the City of Belgrade's strategies and plans and include Belgrade Waterfront, Beton Hall, K-District, Linear park, City on Water, Block 18, and Confluence Park on the Sava and Danube riverbanks and three connections in the form of Ada Bridge, New Sava Bridge, and the Gondola project (Fig. 4.2). These projects mostly belong to Belgrade's wider "natural core"—the confluence of the Sava and Danube, recognized as the basic symbol of its identity.

It is important to mention that the concept of megaprojects is in line with the "post-socialist/proto-capitalist" transition of society and the economy within the capitalist system of neoliberal discourse, characterized by "a poorly developed institutional and democratic culture" (Zeković et al. 2018, p. 156); more on this is presented in Chap. 2. Transition has provided an outline for the use of megaprojects as instruments in the development of Belgrade as a post-socialist city, which is shaped by a mix of economic interests, sociopolitical, and ideological frameworks (Zeković et al. 2018). Some authors (Čukić et al. 2015) have called this investor urbanism, describing it as "practices where the interests of the investor become the main criteria in defining urban growth policies, disregarding any consequences this may have for quality of life and the public interest."

The character of Belgrade's waterfront revitalization will be presented in more detail through several megaprojects, defined in the development strategies and plans for the City of Belgrade. They are characterized by "shiny pictures, big names, high promises of overall general benefit, striking media attention, change of legislation regulating urban planning and construction, with each new project being more ambitious than the last" (Čukić et al. 2015). This research covers ten megaprojects divided into three categories with regard to their character. The first group of megaprojects includes mixed-use urban projects and architectural complexes such as (1) City on Water, (2) Beton Hall, (3) K-District, (4) Belgrade Waterfront, and (5) Block 18. Recreational areas cover two projects—(6) Confluence park and (7) Linear park, while the third group of projects include infrastructural projects that connect the shores (8) Ada Bridge, (9) New Sava Bridge, and (10) Belgrade gondola (Fig. 4.3). These projects mostly belong to Belgrade's wider "natural core"—the confluence of the Sava and Danube. The selected projects will be presented following the same structure and cover the historical background of the project, with a focus on ways in which the site's transformation was seen in earlier

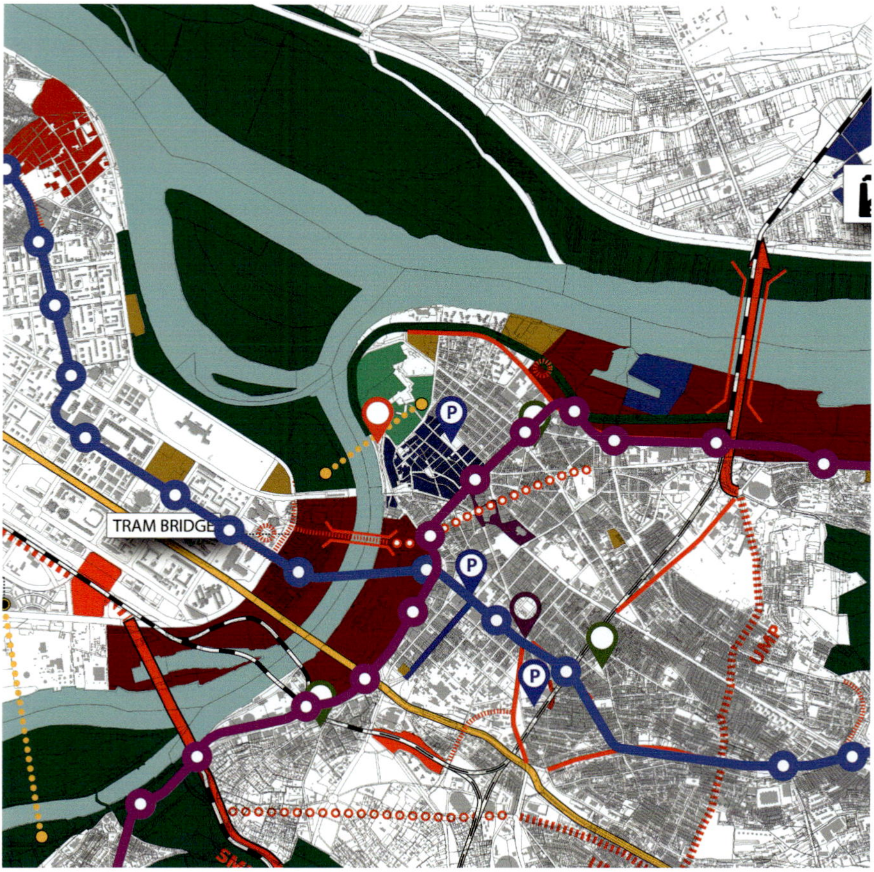

Fig. 4.2 A synthetic suggestion of intervention in the Belgrade coastal zone. Author: Ana Delipara. Archive of the Director of Urban Planning of the City of Belgrade

periods, as well as the contemporary social and political background of the project, the current version of the project and its character, the level of implementation of the project, and the general public and professional attitudes to the project.

4.1 Mixed-Use Projects in Belgrade's Waterfront Area

The City on Water or Belgrade Port Project covers an area on the right bank of the Danube River where the Port of Belgrade is currently located. It was designed by Daniel Libeskind, who defined the face of the urban form, and Jan Gehl, who focused on the quality of public spaces and their connections with the city center. As one of Belgrade's most important projects of the first decade of the twenty-first

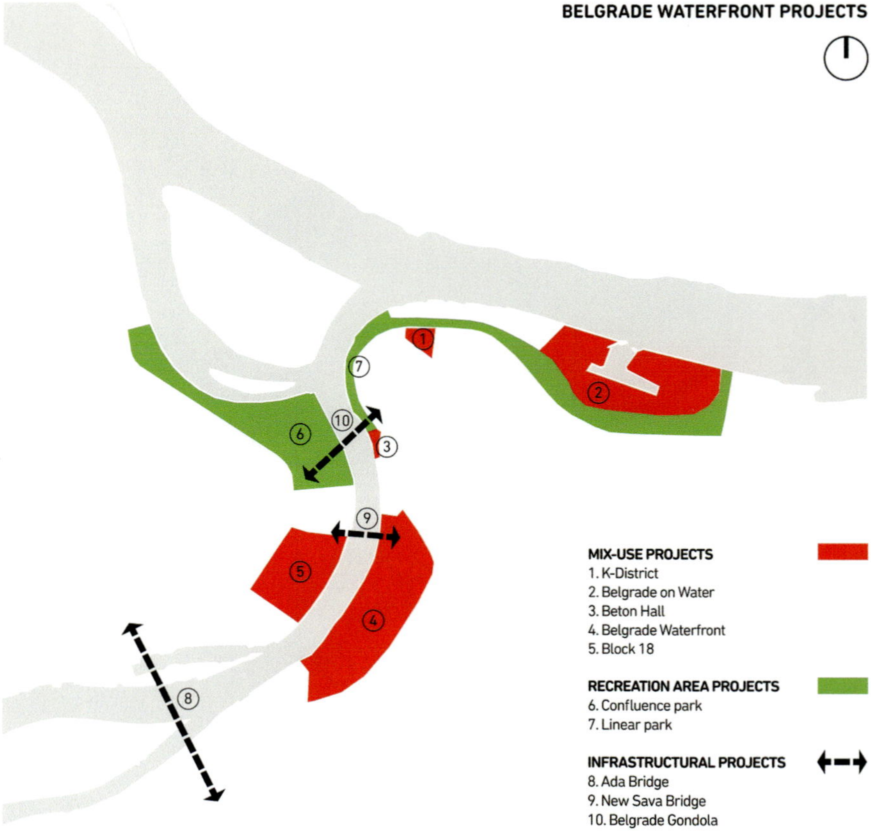

Fig. 4.3 Belgrade Waterfront projects. Vukmirovic 2019

century, it was seen as "an important element in Belgrade's transformation to a European capital with a unique and recognizable identity" (Luka Beograd 2010).

The project was preceded by a spatial-program study carried out by the Urban Institute of Belgrade in the form of an invited competition for five public professional institutions: the Faculty of Architecture of the University of Belgrade, the Association of Urban Planners of Belgrade, the Architectural Association of Belgrade, the Academy of Architecture, and the Architect's Club (Vukmirović 2015). Invited teams delivered five visions for the future transformation of the 470-ha territory that includes the area of the Port of Belgrade. The Urban Institute of Belgrade continued with a Program for the Development of Ada Huja, including the areas alongside the Danube's right bank. The program was followed by changes and amendments to the spatial plan.

The common vision of Libeskind and Gehl Architects (Fig. 4.4) generated a strategic basis for the development of a new urban center that contains an "attractive public space network that gives priority to pedestrians and cyclists, a revitalized

Fig. 4.4 City on Water—Belgrade Port by Daniel Libeskind in collaboration with Jan Gehl. Archive: Belgrade Port

riverbank with public spaces that will improve the appearance and significance not only of the location, but of the whole of Belgrade, a continuous green area along the entire location conceived as a new public park, functions and facilities in an urban space that creates conditions for overlapping various activities, a dynamic atmosphere and openness for different user groups, a variety of districts designed in line with the mixed-use principle, an urban density and architectural accomplishments which would improve Belgrade's appearance and significance" (Luka Beograd 2010). The plan envisaged the construction of residential and commercial buildings, facilities for cultural facilities, a congress center, a school, a nursery, and a hotel. The project's main landmark of the area is a 250-m-high skyscraper, symbolizing the meeting of the two rivers. The City Parliament adopted a Detailed Regulation Plan for an area of 470 ha in Ada Huja in accordance with the Spatial Program for the territory in December 2012.

However, the main goal of the project and its owner was to offer "big investment possibilities to everyone interested in investing in the location" (Libeskind interviewed by Veselinović 2009). Most attention among the media and the public was given to the controversial privatization of the Port of Belgrade and not the quality of the project. "Although a large number of planning prerequisites for the continuation of the project were determined, an unresolved situation with the ownership of land has significantly reduced the interest of the new government elected in May 2014 for this location" (Vukmirović 2015). As a result, the current project status can be understood as frozen.

Beton Hall project was recognized by the General Plan of Belgrade for 2021 and the City of Belgrade Development Strategy (Palgo center 2011) in the domain of Belgrade's brownfield development, as well as the revitalization of Kosančićev Venac and the rehabilitation of Savamala. These documents anticipated the need for an international architectural competition with the main objective to create "a new access point from the capital's riverfront to its historic core and a contemporary architectural anchor point for a vibrant pedestrian zone in one of the oldest and continuously inhabited urban areas" (Beton Hall Waterfront Center 2011). The competition was announced in February 2011, and the results were presented to the public in June 2011.

Five proposals were selected among the 135 submitted competition projects, and two of them were awarded equal first prize. The project *Cloud* designed by Sou Fujimoto Architects attracted the attention of both the domestic and the international public (Vukmirović 2015) because of its attractive appearance and innovative approach in achieving the connection between the riverfront and Kalemegdan (Fig. 4.5). The project was described as brave and innovative in several aspects with the highest emblematic potential among all of the Beton Hall entries, and as an iconic structure which skillfully blurs the difference between the urban and architectural scales and spaces (Beton Hall Waterfront Center 2011).

Even though the work on the Detailed Plan for Beton Hall and the connection between the riverbank and Kalemegdan started in 2012 in accordance with the competition proposal, this project has been almost completely abandoned. It is interesting to add that Beton hall was one of the megaprojects of Belgrade's previous authorities.

However, the basic idea of a direct connection between Belgrade Fortress and the Sava quay has survived and is currently being constructed. It is designed as an

Fig. 4.5 Sou Fujimoto Architects' winning proposal on the Beton Hall Competition. Archive: Belgrade Architect Association

Fig. 4.6 Pedestrian bridge with the sculpture "From there to here." Archive of the Director of Urban Planning of the City of Belgrade

art installation that contains a contemporary sculpture and a pedestrian bridge. The sculpture entitled "From there to here" is the work of Richard Deacon and Mrdjan Bajić, while the pedestrian bridge was designed by Branislav Mitrović in collaboration with the Traffic Institute CIP. The bridge (Fig. 4.6) is made from concrete and glass and includes an elevator, a staircase, a passageway, and a circular ramp which is an integral part of the sculpture made from concrete and steel.

K-District megaproject, also known as The Beko Master Plan, is an urban regeneration proposal for a multifunctional complex, displacing the former "Beko" textile factory in the immediate vicinity of the Belgrade Fortress. The basics for the project can be found in the Master Plan of Belgrade 2021, which states that the block needs to be modified in accordance with the character of the city center and that the earlier detailed plan is not mandatory unless the changes are in accordance with the new plan. In addition, in the Development Strategy of the City of Belgrade, the project is seen as one of the priority projects defined as the "Belgrade brownfield" areas (Ada Huja, the Shipyard, Beton Hall, etc.).

The Beko complex project was presented as *the city's new and happening center* for the first time at Belgrade Design Week in 2012. It foresees the covering of 94,000 m^2 and includes cutting edge residential spaces, galleries, offices, a five-star hotel, a state-of-the-art congress center, retail spaces and a department store, etc., which will infuse a new life to the historical quarter of Dorćol and form a destination point between Belgrade Port, Beton Hall, and Savamala. The initial design was made by Zaha Hadid Architects, and it was conceived as a mark of the continuance of Belgrade's "modernist" movement, which was abruptly discontinued in the 1980s (Slavkovic 2014).

Fig. 4.7 K-District. Project designed by Boris Podrecca. Archive of the Director of Urban Planning of the City of Belgrade

In 2017, the Lamda Development Company sold the location to a new Dutch investor firm Pinnacle (Mučibabić 2018a). The Dutch company decided against the solution proposed by ZHA, and the job was entrusted to Boris Podrecca's studio from Vienna. They started the work on the K-District project at the beginning of 2018. Podrecca designed a residential district with different types of housing, upgraded with complimentary facilities and open spaces, which crosses a 35-m boulevard, and its shape and height do not jeopardize the view from the fort to the Danube (Fig. 4.7). Although the local government announced the beginning of the implementation of this project in mid-2018, this has not yet happened, so this project still remains on paper and in newspaper articles.

The Belgrade Waterfront is an ongoing urban renewal project situated on the right bank of the Sava River, between the Belgrade Fairground and the protected area of Kosančićev Venac. The area, known as the Sava Amphitheater, also includes the location on the left bank (known as Block 18 at present), and it has been the subject of a series of development initiatives in the past. During the late 1970s, it was considered in the Study for the City Centre—Sava Amphitheater (Jovanović and Perović 1975; Glogorijević and Graovac 2018), known as the Project for Friendship Home (Djukić et al. 2014). Bearing in mind the past activities and the strategic character of the area, the location is recorded as a development area in different ways: as the Sava Amphitheater in planning documents; as part of New Belgrade in the General Urban Plan of Belgrade 2021; and as the Sava waterfront development in the City of Belgrade Development Strategy. Accordingly, the Urban Institute of Belgrade began to prepare an urban study as a prerequisite to the announcement of an international competition for the Sava Amphitheater.

However, the planned international competition was not announced, since a new initiative for this location appeared. The competition was preceded by national and local elections in 2012, in which the project was a key element in the political campaign by the Serbian Progressive Party, which won the elections. Driven by a desire to restart the economic development of the state, megaprojects like this were seen as the backbone of the change. In order to avoid the failure of the previous initiatives for the reconstruction of the Sava Amphitheater, the project was designed as a public–private partnership. Thus, negotiations with UAE investors began, after which the Belgrade Waterfront Master Plan was presented to the international public in Dubai in March 2014. Soon after that, the plan was presented to the local public.

Belgrade Waterfront, also named the "Dubai of Europe," referring to its investors from the Emirates, or "Belgrade's Manhattan" (Folić in Roth and Dudas 2015), is seen as an initiative that will create jobs and growth and turn Belgrade into a business hub for the Western Balkans. The government gave the project the status of national importance, which has allowed the acceleration of the process of defining the framework for its implementation (Vukmirović 2015). Over a period of several months, the Spatial Plan for Belgrade 2021 was changed and the Spatial Plan for the Development of the Coastal Area of Belgrade's riverside was adopted for the Belgrade Waterfront project. This was "a kind of precedent since it was about the central city territory" (Glogorijević and Graovac 2018). From the presentation of the project initiative to the beginning of its implementation, the project has followed a series of controversial and unexpected events that have influenced its overall reputation among the members of the professional public (Zeković et al. 2018; Čukić et al. 2015; Roth and Dudas 2015). These include adapting urban planning laws to suit the needs of the Belgrade Waterfront project, having a non-transparent contract with partners from UEA, i.e., with the firm Eagle Hills, and the violent demolition of a certain number of objects on site during the night in 2016.

With these situations in mind, Belgrade Waterfront project could also be described as "the product of a top-down central planning process" of the "economically liberal, socially conservative, proudly nationalist" government that "is building an architectural monument to the new Serbia" (Roth and Dudas 2015). "The monument" will represent the values that the government wants to project, such as "technocratic, responsibly elitist, supply-side and foreign-investor focused, tourist-friendly" (Folić in Roth and Dudas 2015) and the government's desire to take "urban renewal to new heights—a future smart city that combines commerce, culture and community" (Belgrade Waterfront 2014).

Belgrade Waterfront project envisions the construction of offices and luxury apartment buildings, eight hotels, a shopping mall, and a 250-m-high tower as the main symbol both of the area and also of Belgrade (Fig. 4.8). From the investor's perspective, it will become "a truly mixed-use environment that will usher in a new era of prosperity for the Serbian capital" (Belgrade Waterfront 2014). But at the same time, some authors are wondering: why luxury apartments are being built when "thousands of people are without a permanent housing solution"; why the largest shopping mall in the Balkans is being constructed when "each day the amount of people living below the poverty line increases"; and why the new retail and office spaces are being developed when "'for rent' signs have continued to fade for years on buildings of the same kind" (Čukić et al. 2015).

The implementation of the project started in 2014. So far, part of the Sava Promenade and a double tower have been completed, while the planned shopping center, BW Gallery, Belgrade tower, two solitaires, and part of the street network

Fig. 4.8 Belgrade Waterfront. Archive: Office of the Director of Urban Planning of the City of Belgrade

are currently being constructed. Although it is still accompanied by negative opinions shared by experts and the general public, this is a project that characterizes contemporary trends in the construction of cities, based on liberal capitalism and investor urbanism. For this reason, the real situation regarding the transformation of the Sava Amphitheater, as well as the functioning of Belgrade, will be known in a few years. "Should it succeed, the project will remake the city more profoundly than anything since World War II" (Roth and Dudas 2015).

Block 18 is located within the central zone of Belgrade, more precisely in the Sava Amphitheater, on the left bank of the Sava and it thus makes direct contact between the urban tissue and the river. The first ideas for this area appeared after WWI in the General Regulation Plan from 1923, within which the author, Djordje Kovaljevski, produced a supplementary document entitled "An illustrative plan of the city on the left bank of the Sava", although Zemun had not yet merged with Belgrade.

This idea has since inspired other initiatives for the urbanization of this area. Zemun was annexed to Belgrade in 1934, with a direct connection being established by the construction of King Alexander Bridge. In 1938, work started on filling the terrain between the Belgrade–Zemun road and the Danube Quay, while the fairground was built in the immediate vicinity of King Alexander Bridge, now known as the Old Fairground. After WWII several planning documents and studies (General urban plans 1950, 1972, 1985, 2003, plans and studies for the central zone of New Belgrade from 1975, 1981) dealt with this territory, but until the present day, most of the space has been in extremely poor condition, although it is located on one of the most valuable city locations.

Its position in relation to the Sava River and the center of New Belgrade makes this area the most important location for integrating the two coasts, as well as being a link between the existing and planned developments. Also, this area belongs to a part of the Sava riverbank known as Sava Quay developed in 1967 according to a project by Branislav Jovina, who received the October Award in 1969 for this project. This promenade is therefore under the previous protection of the city unit of the Coastal Zone of New Belgrade. However, at present, most of Block 18 is made up of family homes, predominantly low density and poor in quality. It is characterized by a relatively old housing fund and a spontaneous network of local streets. It also needs to be mentioned that the location of Block 18 has never previously been analyzed without the area across the river, which is now the area of the Belgrade Waterfront project. These two locations have always been planned and studied integrally (Perović, Jovin, etc.), and now for the first time, by adopting the Spatial plan for special purposes in 2014, there is a need for a proposal for just one location, while the other has already been given a solution by a private investor (Belgrade Waterfront).

Therefore, an international urban architectural competition was announced in March 2016 to find a proposal that would define the morphology of the newly planned urban structure and its relationship with the environment, the Sava, and its opposite coast (Društvo arhitekata Beograda 2016). Eighteen works were received in the competition, and the winning project designed by Panic and his team proposed a new urban complex organized as a pedestrian area with plenty of biking trails and streets in which vehicle access would be allowed only to delivery vehicles (Fig. 4.9). It defined two towers along Vladimira Popovica Street of 100 m in height, incorporated into the existing silhouette of New Belgrade, which will not compete with the future Belgrade Tower.

Although it envisages the construction of a concert hall and contemporary city gallery, the winning urban design solution is not defined in the contemporary spirit. The solution is reduced and modest, and it does not correspond with the matrix of a modern city, which is what New Belgrade is, and the architectural form and appearance of the buildings seem to be outdated. Some solutions from the 1970s and 1980s seem much more contemporary than this winning proposal. The physical continuity with New Belgrade's urban matrix, as well as the connection with the Belgrade Waterfront project, is achieved by the pedestrian bridge, which is an

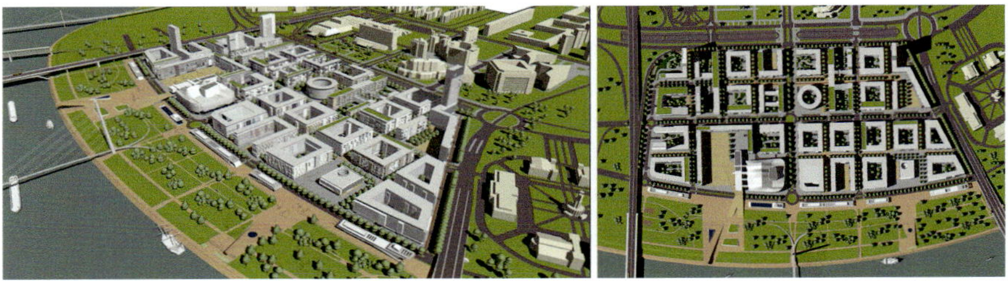

Fig. 4.9 Wining proposal for Block 18. Authors: Panić et al. (2016). Archive of the Director of Urban Planning of the City of Belgrade

extension of Zoran Djindjić Boulevard. This project will be the basis for the preparation of a detailed regulation plan for Block 18, while other activities in this area are not yet known.

4.2 Recreation Area Projects Along the Waterfront

Confluence or **Friendship Park** (Park prijateljstva) is situated in New Belgrade and it is one of the largest green areas in the city. The first idea for this park as Friendship Park, with a memorial character, was announced in honor of the First Conference of non-aligned countries in Belgrade in 1961, which was elaborated within the 1962 Regulatory Plan of New Belgrade. It followed the basic idea of building New Belgrade on the principles of the modern movement. A special segment of this plan referred to the arrangement of recreational and green surfaces, both in terms of their capacity and their spatial experience. Green areas were planned for residential areas and in recreation zones, alongside the SIV building, CK SKJ, at faculties and hospitals, as industrial and protective greenery (belt between highway and industry), as well as greenery in the form of lawns and greenery with special purposes. In the regulatory plan, recreational complexes were foreseen in housing areas and along the line of the coastal area. Urban parks with different characters were planned for the area around the SIV building (the Palace of Serbia), such as Friendship Park (covering an area of 12 ha), the Park of Art (19.5 ha), Sava Park (on the coast of the Sava River, 19 ha), and the Amusement Park (next to the railway station in New Belgrade, 10.5 ha) (Vesković 2011).

The final solution for the appearance and organization of the park area was shown in the project called Potez, which won the first prize in the Yugoslav competition in 1965. It was a challenging competition topic because it dealt with the field of horticulture, which is not often a competition task, as well as the location of the park among the most important state buildings. Nineteen projects entered the competition, three of which were awarded prizes, and eleven of which were purchased. Milan Pališaski was the winner of the competition and the construction of the park began in accordance with his proposal. The memorial park has never been fully completed in the way that it was conceived, but nevertheless today it is a significant and valuable part of the historical heritage of the city. It is also called "a living monument" because, since 1961, when the Aleja of Peace was formed, over 200 trees have been planted along the promenade paths and lateral alleys. For decades, presidents, celebrities, and officials from all continents have planted trees as a symbol of friendship, such as US Presidents Gerald Ford, Jimmy Carter and Richard Nixon, Presidents of the Soviet Union Leonid Brezhnev and Mikhail Gorbachev, East and West German Presidents Erich Honecker and Willie Brandt, Indian Prime Ministers Jawaharlal Nehru and Indira Gandhi, King of Norway Olaf Peti, Iranian chess player Reza Mohammad Pahlavi, English Queen Elizabeth II, Swedish King Karl XVI Gustaf, Cuban President Fidel Castro, Egyptian President Gamal Abdel Naser, and North Korean leader Kim Il Sung.

The current state of this cultural asset is very poor. The stone slabs on which the names of people are written are completely covered with grass, the gardens are unregulated, the pedestrian paths are damaged, and since there is no illumination, most people avoid it in the evening. Due to all of the above, there was a need to stop the further deterioration of this area and to reconstruct it in order to exploit its full potential. For this purpose, Gehl architects carried out a study and suggested a concept for the future revitalization and arrangement of this space. It's about 91 ha in size and it extends from Branko's Bridge to the hotel Yugoslavia and includes Art and Friendship parks with the Aleja of peace. The park is designed to be used by all citizens. The new content will be refined, attracting visitors and allowing them to have fun, rest, and feel safer.

In accordance with this, Gehl architects created the vision called "Belgrade Confluence Park: People's Park", defining it as a lively and dynamic park that invites people to spend time there. It will be an accessible and inviting place for Belgradians and local people of all ages, and a vibrant local space where visitors and tourists are attracted to the authentic atmosphere among the locals, where they can enjoy the local food, events, and nature as a part of their stay in Belgrade. An Aquatorium, Ferris Wheel, Natural museum, Music hall, and the new connection with the old core of Belgrade (gondola) are the main attractors of Confluence park (Fig. 4.10). The positions of these facilities will be determined by an international urban architectural competition, defined by the Urban plan for the regulation of multifunctional sports and cultural content in Friendship Park—Confluence park, produced for this area by the urban planning institute for Belgrade.

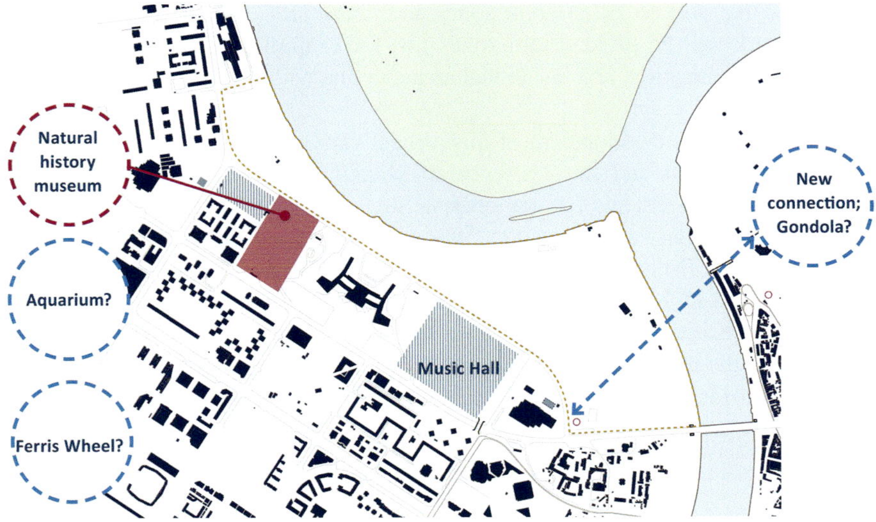

Fig. 4.10 Belgrade Confluence Park concept. Authors: Gehl Architects. Archive of the Director of Urban Planning of the City of Belgrade

So far, one part of this project has been completed. It is located in the zone of the Museum of Contemporary Art and is a path called "Art", one of the seven established thematic routes to the river. This part of the park has revived since its reconstruction, because it contains interactive playgrounds and a path leading to the Museum of Contemporary Art which, after years of reconstruction, has reopened. It remains to be seen how the remaining phases of the reconstruction of this area will take place, since they require more serious interventions, and also how the space will be maintained.

Linear Park is planned along the abandoned railway line, following the right coastline of the Sava and Danube Rivers for a length of a few kilometers from Beton Hall to Belgrade port. Belgrade's train network and its organization represent a decades-old issue that has captured the attention of Belgrade's urban planners. Several scenarios for addressing this issue appeared. They include the position of the main railway station, but also its connection with the Port of Belgrade, located on the Danube, the relocation of the port, the construction of the metro system, and the development of the area within the Sava Amphitheater. Since these are the most important long-term projects of the city of Belgrade, their final determination has resulted in a number of other projects concerning lowering Belgrade to the rivers.

The move of the railway linking the main railway station with the port of Belgrade was, until recently, an undefined topic, because it had to be determined by when the existing railroad would be displaced. The Belgrade Waterfront project and the plan for the Belgrade port area, defined within the City on Water project, have provided the basis for the future treatment of this area, which has been a decades-long physical barrier to establishing direct links with the Danube coast (Mučibabić 2018b). Keeping in mind the form, character, and dimensions of this area, basic concepts are currently being developed to transform this space into a linear park that will cover different zones and ambiences. In accordance with this idea, the park will be divided into zones for, for example, nightlife, sport, recreation, culture, education, and residential areas, in line with the main character of the contact areas.

The status of the development of this project varies according to the part of the city being observed and the existence of planning and project documentation. Works on the arrangement of Karadjordjeva Street and a part of the Sava Quay near Beton Hall are underway (Fig. 4.11). Work will be carried out on the basis of the project design by Boris Podrecca and the CreativeBGD team. The project envisages the construction of Karadjordjeva Street in accordance with the green traffic hierarchy, while facilities are planned along the river to make longer stays outdoors possible. For the rest of the area, the Urban Planning Institute is working on a plan that will consider all the possibilities and potential and provide guidelines for further development and regulation.

Fig. 4.11 Karadjordjeva Street. Authors: CreativeBGD, 2016. Archive of the Director of Urban Planning of the City of Belgrade

4.3 Infrastructural Projects

The Ada Bridge is an integral part of the Inner City Semi-Ring Road directed around the city, the central parts of New Belgrade and Zemun. The preconditions for its construction can be found in the General Plan of Belgrade for 2021 (2003) and the City of Belgrade Development Strategy (Palgo center 2011). In accordance with these documents, Belgrade Urban Planning Institute developed a Detailed Regulation Plan for the bridge, adopted in September 2007.

Following the recommendations of the General Plan of Belgrade for 2021, an international competition for a design proposal was launched in 2004. Eleven proposals were selected, among which the first prize was given to Slovenian design team led by Viktor Markelj and Peter Gabrijelcic. The project for the bridge was developed in 2006, and work started on it in 2008, while the bridge was opened to traffic in 2012 (Fig. 4.12).

Ada Bridge was initially seen as the future landmark of the City of Belgrade, since its main visual characteristic is a 207-m-high pylon that enables two asymmetrical spans to cross the river. "The 920 m long connection of Sava's riverbanks has six lanes for road vehicles, two light rail railway tracks, and two pedestrian and cycling paths" (Markel and Gabrijelčič 2012), and it is recognized as "the largest bridge in the world suspended by just one pylon."

Fig. 4.12 Ada Bridge. Archive: Office of the Director of Urban Planning of the City of Belgrade

Although it opened to traffic in 2012, certain parts of this project have only recently been completed, such as the establishment of a tram network on this corridor. However, some controversies that relate to the completion of this project concern the pylon itself. In a direct conversation with Prof. Gabrijelcic in 2014, the author found out that the top of the pylon is still missing. However, although strongly criticized for the total material value and nontransparency of its implementation (Vukmirović 2015), the bridge has indeed become one of the key landmarks in the overall image of the city, which was one of the primary objectives of the project.

Within the strategy of "Belgrade returning to its rivers" and the development goals of the Belgrade Waterfront and Block 18 projects, there was a need to build **a new bridge on the Sava River**. Based on the locations of these two big projects and the character and capacity of the existing and planned street network, the best position for this bridge was defined as the location of the existing tram bridge on the Sava. The Belgrade Land Agency has made several proposals related to the capacities of the future bridge, while the solution for the future form and layout of the bridge was proposed by the Traffic institute CIP in the competition for the New Sava bridge design. The new bridge will have an attractive, but not such an advanced form, with two touching arches (Fig. 4.13). The proposal will repeat the recognizable silhouette element of the old bridge, with its arched construction, in order to save its memory, while its functionality will enable a greater flow of traffic. Considering the functional aspect, it will have pedestrian and cycling paths, two automobile lanes in both directions and a lane for trams.

Although it can be considered rational in relation to the current and future needs of the city, citizens have not responded positively to this idea. The reasons for this are primarily reflected in the need to relocate the existing bridge, which is a very important element of the image of the city of Belgrade.

Belgrade Gondola is a project that the Belgrade local authorities justify by the need to form a direct connection between Confluence Park and Belgrade Fortress.

Fig. 4.13 New Sava Bridge. Author: Traffic Institute CIP. Archive: Office of the Director of Urban Planning of the City of Belgrade

Although it seems so, the idea of this project is not a new one. According to Stojanovic (2018), an idea came up in the period between the First and Second World War, entitled "flying tram", that would connect Belgrade with Zemun. During the 90s of the twentieth century, the gondola again became a possibility, but at a different location, with the aim of connecting Kosutnjak and Banovo Brdo with Ada Ciganlija. This idea was further developed in 2012 by the Juginus Institute, when an extension of the planned line to New Belgrade's Block 44 was proposed.

However, the idea for the gondola in the city center came about again, led by the will of Belgrade's governing authorities to improve its tourist offer by advancing its public transport and make it more attractive. This time a new location was considered, in which the gondola would connect Kalemegdan with Confluence Park. Although the study "Belgrade Confluence Park: People's Park" considers this form of connection between the two parts of the city an inadequate solution, characterizing it as "expensive to run and maintain, losing out on the attraction of sensing the river by flying up high, and inconvenient for people to use (have to wait, not always available when needed, etc.), and not for everybody because tickets are charged" (Gehl Architects 2017), further work on this project continues.

The detailed regular plan for the Gondola was produced by the Institute of Architecture and Urbanism of Serbia in 2018 (Stojanovic 2018). Based on the study that preceded the development of the plan, it was concluded that Belgrade needed a gondola at this location. Arguments in its favor are: the height is determined by the waterway; it is an acceptable investment; it is a tourist attraction; and it is not possible to form a different form of connection. The gondola will extend over a kilometer in length, 300 m above the river. This transport system will have two stations (one near Belgrade fortress and another near the Ušće shopping mall) and six decorated pillars, two next to stations at a height of 15 m. With a speed of 5–6 m/s, it will transport 3,000 passengers per hour in both directions. It was presented to the public as an extremely important project, although most of the professional public disagrees with this decision.

The extremely negative attitude of the wider and professional public toward this project came after 150 trees were cut down as preparatory works on the project. Two main reasons were given for the negative attitude, the first being the number of trees cut down and the violent behavior toward nature, and the second being the illegal start of works on the project as it was not licensed. This initiative has grown into a petition signed by more than 15,000 citizens, which points to procedural irregularities into the hope of preventing the realization of the project. The irregularities are related to the detailed regulation plan for the Kalemegdan–Ušće gondola and the inadequate conditions and measures of protection issued as defined by the Republic Institute for the Protection of Cultural Monuments. It is currently not known what will happen to the implementation of the Belgrade Gondola project.

4.4 Conclusion

All the megaprojects planned in the Sava and Danube waterfront area have been presented to the public as something that will be completed in the near future, within a couple of years. However, this can only be said for the Belgrade Waterfront project and the small part of the Confluence Park project that has already been constructed (see Table 4.1). Other projects require more reflection before their implementation, while yet others will wait for future times.

Table 4.1 Actual implementation of Belgrade's megaprojects

Main character of the projects	Name of the project	Investors	Character of the project	Status of the project
Multi-use urban areas and architectural complexes	City on Water	Unknown	Luxury apartments Luxury hotels Shopping mall Offices Culture Public spaces	Frozen
	Beton Hall	Unknown	Shopping spaces Office spaces Public spaces	Replaced by a smaller project

(continued)

Table 4.1 (continued)

Main character of the projects	Name of the project	Investors	Character of the project	Status of the project
	K-District	Pinnacle (Holland)	Luxury apartments Office spaces Retail spaces Public spaces	On paper only
	Belgrade Waterfront	Eagle Hills (UAE)	Belgrade tower Luxury apartments Luxury hotels Shopping mall Offices Culture Public spaces	Under implementation
	Block 18	Unknown	Residential areas Business Culture Hospitality Public spaces	Under implementation based on the winning competition proposal
Recreation areas	Confluence Park	City	Green areas Playgrounds Public spaces	Under implementation
	Linear Park	Unknown	Sport and recreational area Green areas Playgrounds Public spaces	Project initiative
Infrastructural projects	Ada Bridge	City	Bridge with access roads	Completed
	New Sava Bridge	Unknown	Bridge	Project development
	Belgrade Gondola	City, Forests of Serbia	Gondola	Unknown

However, these projects show that Belgrade does not think comprehensively about its identity and image, although this is defined in its planning documents. There have always been hidden motives; previously they were state ambitions, and now the motive is profit, justified by planning documentation. The period of a strong state with strong institutions is behind us, and it is possible to see the outcomes of its major projects. What will happen after the period we are witnessing, we can only assume, and the witnesses will be new generations. The Sava and the Danube, I hope, will be still here, as an inseparable part of this city.

References

Belgrade Waterfront (2014) The wave of new life. Belgrade. Belgrade Waterfront. (Online) http://www.belgradewaterfront.com/Belgrade-Waterfront/116/Wave-of-New-Life.shtml

Beton Hall Waterfront Center (2011) Beton Hall Competition. (Online) http://www.betonhala.com/2011/

Čukić I, Sekulić D, Slavković L, Vilenica A (2015) Report from Belgrade Waterfront in Eurozine. https://www.eurozine.com/report-from-belgrade-waterfront/. Accessed online 5 July 2019

Djukić A, Vukmirović M, Vaništa Lazarevic E (2014) Planski okvir, projekti, konkursna rešenja i vizije za uređenje Savskog amfiteatra (The planning framework, projects, competition proposals and visions for the regulation of the Sava amphitheater). Izgradnja 68(3/4):103–120

Društvo arhitekata Beograda (2016) Konkurs za blok 18 na Novom Beogradu (Competition for the Block 18 in New Belgrade). Beograd. Društvo arhitekata Beograda

Gehl Architects (2017) Ušće Park—People's Park. Gehl Architects, Copenhagen

Generalni urbanistički plan 2021 (General Urban Plan of Belgrade 2021) (2003) Beograd. Urbanistički zavod Beograda

Glogorijević Ž, Graovac A (eds) (2018) 70 Godina Urbanističkog zavoda Beograda Knjiga II: Planovi (70 Years of Belgrade Urban Planning Institute). Beograd. Urbanistički zavod Beograda

Jovanović S, Perović M (1975) Gradski centar - Savski amfiteatar (City center—Sava's Amphitheatre). Beograd. Urbanistički zavod Beograda

Luka Beograd (2010) City on water masterplan. (Online) http://www.lukabeograd.com/en/city-on-wather/master-plan.html

Markel V, Gabrijelčič P (2012) The Ada bridge/Architectura d.o.o. Archidaily (29 Feb 2012). (Online) http://www.archdaily.com/?p=212085

Mučibabić D (2018a) "K distrikt" umesto "Beka" ("K-District" instead of "Beko"). Politika 12.02.2018

Mučibabić D (2018b) Linijski park po uzoru na njujorški Haj lajn (Linear park by model of New York's High Line). Politika 06.03.2018

Palgo centar (2011) Strategija razvoja grada Beograda (City of Belgrade development strategy). Beograd. Grad Beograd

RAREI and Palgo centar (2017) Strategija razvoja grada Beograda: Strateški ciljevi, prioriteti i mere održivog razvoja do 2021 (City of Belgrade Development Strategy). Beograd. Grad Beograd

Regionalni prostorni plan administrativnog područja Grada Beograda (Regional plan for the administrative territory of the City of Belgrade) (2004) Beograd. Urbanistički zavod Beograda

Republička agencija za prostorno planiranje (2014) Prostorni plan područja posebne namene uređenja dela priobalja grada Beograda - područje priobalja reke Save za projekat "Beograd na vodi" (Spatial plan for the development of the special purpose coastal area of Belgrade—the riverside area for the "Belgrade Waterfront" project). Beograd. Republička agencija za prostorno planiranje

Roth M, Dudas K (2015) History of the present: Belgrade in Places Journal. https://doi.org/10.22269/151005. Accessed 19 July 2019

Slavkovic Lj (2014) U potrazi za transparencijom. Megdan oko beogradskog projekta studija Zahe Hadid. Super Prostor (Megdan around Belgrade project study by Zaha Hadid). (Online) http://www.superprostor.com/megdan-oko-beogradskog-projekta-studija-zahe-hadid/10606

Stojanovic M (2018) Hoce li Beograd (konacno) dobiti gondolu? (Will Belgrade (finally) get its gondola?) Published online in Gradnja.rs on 18th October 2018

Studija beogradskog priobalja I faza (The Study of the Belgrade coastal area) (2007) Beograd. Urbanistički zavod Beograda

Veselinović I (2009) Famous architect Libeskind is designing the "City on Water" in Belgrade. Interview with Daniel Libeskind. ProMoney (5)

Vesković I (2011) Park prijateljstva u Novom Beogradu (Friendship park in New Belgrade). Nasledje (12):203–216

Vučelić V (1977) Oko reke i o rekama (Around and about the rivers). Urbanizam Beograda. IX:10–14

Vukmirović M (2015) Belgrade: the quest for the desired city image. In: Planning capital cities: Belgrade, Bucharest, Sofia. Verlag der Technischen Universitat Graz, Graz, pp 188–211

Zeković S, Maričić T, Vujošević M (2018) Megaprojects as an instrument of urban planning and development: example of Belgrade Waterfront. In: Hostettler S, Najih Besson S, Bolay JC (eds) Technologies for development. UNESCO 2016. Springer, Cham

Chapter 5
Reimaging Belgrade: The Case of Savamala District

Biljana Arandelovic

Abstract Belgrade is now becoming one of the most vibrant Southeast European urban centers, and it is attracting many tourists, mostly among the younger population, due to its hedonistic alternative culture and club scene reputation. Belgrade's Savamala district has transformed over the last few years into an energetic creative urban district where the old buildings in this historical neighborhood are gaining a new role of reimaging Belgrade. The cultural renaissance of Savamala began in 2007 and transformed it into a creative district through various informal research activities, small-scale cultural projects, and creative clusters. The very first activity of this kind was *Magacin* cultural space in Kraljevića Marka Street, which opened in 2007. This continued with CK Grad in 2009 and with the founding of Mixer House in 2012. Another important project that has influenced the development of Savamala as a creative district is the Urban Incubator Belgrade project, initiated in 2012 by the Goethe-Institute Belgrade. After the Belgrade Waterfront project was announced in 2014, changes in Savamala have intensified. Savamala is in the process of deep change: on one hand, it has the building site of Belgrade Waterfront, and on the other hand, culture venues which are the result of a long process of creative and culture-led urban development. This chapter discusses the role of Savamala in reimaging Belgrade, while positioning the district as the center of Belgrade's new urban renaissance.

Keywords Savamala · Creative city · Urban development · Place branding · City branding · Creative economy · Creative cluster · Culture-led urban development

5.1 Development of Savamala

Savamala district is a part of the old town with extraordinary spatial potential; it is located about 0.7 km from Terazije Square and around 1.3 km from Republic Square, two of the most important squares in the historical city center. The district stretches along the Sava River and is rich in tradition, history, and heritage. It is difficult to determine its exact boundaries. Savamala district was an important area in oriental Belgrade, located around the present-day Sava Wharf on the southern

© Springer Nature Switzerland AG 2020

B. Arandelovic and M. Vukmirovic, *Belgrade*, The Urban Book Series,
https://doi.org/10.1007/978-3-030-35070-3_5

bank of the Sava. Its present-day streets that show traces of oriental Belgrade are 7th July Street, Gračanička, Car Lazar, Karadjordjeva (to the Bristol Hotel), Braća Krsmanović Street, and Kraljević Marko Street (Djurić-Zamolo 1977) (Figs. 5.1, 5.2, 5.3, and 5.4). The center of Savamala district is considered to be the location of the former Little Market (Mali pijac), which is now part of the square in front of the Belgrade Cooperative building. The Little Market stimulated the building activity in Karadjordjeva Street, which started in 1907 (Pavlović-Lončarski 2005).

The name Savamala comes from the name of the Sava River and the Turkish word for a district or a block of houses, *mahala*. It lies south of Kalemegdan Fortress and was actually Belgrade's first district that was constructed outside of its walls. For this reason, this district can be considered the cradle of modern Belgrade, despite Turkish garrisons still being present in Kalemegdan Fortress at the time of its origin in the early nineteenth century (Popović 1997). Although located right next to the historic core of Belgrade, the area of Savamala district has long remained uninhabited. The main reason for this is that part of it was covered in a swampy area, which was a result of the spring flooding of the Sava River. On maps of Belgrade from the period of Prince Eugene of Savoy (1663–1736), three main routes that go from Belgrade Fortress can be noticed. One goes along the river Danube, which later became Dušanova Street, the second starts from the middle gate and forks into two parts (the first has become one of Belgrade's main arteries—King Milan Street, the other is the widest and longest street in Belgrade—King Alexander Boulevard), and the third route passes by the Sava River—today's Karadjordjeva Street.

The Savamala district area has historical importance. During the First Serbian Uprising (Meriage 1978) Serbian rebels conquered Belgrade, taking it back from the Ottomans from their base in Savamala, under the leadership of Karadjordje in 1806.

After the failure of the uprising, Belgrade again became Ottoman in 1813. In 1815, Prince Miloš Obrenović entered into an agreement with the Ottomans and received this terrain as a gift. Soon after this, he raised a small settlement on the Sava coast, which was the first time in the long history of Belgrade that the city had moved to the Sava's slope. Until then, it had developed on the Danube slope.

Savamala's official urban development started after 1830 by the order of Miloš Obrenović, the Prince of Serbia and founder of the Obrenović dynasty. 1830 was the year when a Hatti-sherif was issued (a proclamation by an Ottoman Sultan) to give the Serbian government greater autonomy from Turkey. Prince Miloš Obrenović ordered a development plan for Savamala, which in addition to the regulation of the land included the relocation of its inhabitants, all in order to connect this area with Belgrade's existing municipalities (Pavlović 2013). The cemetery had already been moved in 1828. The major change was the demolition of the old ruined houses and the building of new ones, which happened after the new regulation plan of 1834. The old houses were demolished, and some of the empty lots were taken by the state. Prince Miloš Obrenović had a property right under Ottoman law, while part of the land was divided among wealthy merchants and state officials (Petrović 2010).

Prince Obrenović was aware of the importance of proximity to the river and aimed to transform Savamala district into the center of the economic, social, and cultural life of Belgrade, giving Karadjordjeva Street special importance and

Fig. 5.1 Kraljevića Marka Street, Savamala. Photographer Biljana Arandelovic. Rights reserved

Fig. 5.2 Gavrila Principa Street, Savamala. Photographer Biljana Arandelovic. Rights reserved

Fig. 5.3 Gavrila Principa Street, Savamala. Photographer Biljana Arandelovic. Rights reserved

Fig. 5.4 Gavrila Principa Street, Savamala. Photographer Biljana Arandelovic. Rights reserved

designating it as one of the most significant streets in the city at that time. It is evident that the European urban model was the inspiration for its construction and development. Karadjordjeva Street, the biggest street in Savamala, which runs along the riverfront, was once lined with numerous famous hotels, restaurants, and shops. During a large part of the twentieth century, a period when the whole district was neglected, the purpose of Karadjordjeva Street changed, and it became the main arterial road for international traffic through Belgrade, frequently being used by heavy trucks that brought noise and pollution to Savamala (Figs. 5.5, 5.6, and 5.7). This situation is currently in the process of change, and work on the long-awaited reconstruction is underway in the area from Brankov Bridge to Paris Street, with early 2020 as the completion deadline.

The most important period in Savamala district's development was during the nineteenth century, when this area became the home of small traders and artisans, all positioned around the harbor. This transformed the area into an elite district with representative administrative buildings and luxury homes. Among the very important changes along Karadjordjeva Street, carried out on the order of Prince Miloš Obrenović, was the construction of Djumurkana, the public building that served as the center of cultural life in Belgrade at the time. This was the first permanent Serbian theater, which worked from 1841 to 1842. The port in Savamala district had an important role as an international trade center. A banking and insurance joint-stock company was established in 1882 in Karadjordjeva Street, the Belgrade Cooperative Building, in one of the most beautifully ornamented

Fig. 5.5 Karadjordjeva Street, Savamala. Photographer Biljana Arandelovic. Rights reserved

buildings in the city. The biggest construction activity in Belgrade at the end of the nineteenth century and at the beginning of the twentieth century took place in the area around the port and spread to Karadjordjeva Street. That was an important period for Belgrade's development, and Savamala district was at the center of it (Fig. 5.8). Savamala's restaurants and bars, mainly in Karadjordjeva Street, had a major role in the development of its social and cultural life as they were often a stage for concerts and performances. This atmosphere attracted merchants and rich people to Savamala, both to open their businesses and to move there as well (Fig. 5.9). Soon, Savamala district became a business center and was the area where Belgrade's first bourgeois class was created. Its favorable position near the Sava port allowed good connections with Europe, after a couple of centuries of Turkish oriental dominance (Petrović 2010). Beautiful richly decorated historical buildings can still be found in this area, exposing a range of facades, as a testimony to Savamala's past glory and the rich architectural and cultural heritage of the district, for example, Academism, Art Nouveau, and Secession architecture, but also Neo-Renaissance and modernist details. Unfortunately, some of these historical buildings have deteriorated because of inadequate maintenance, often due to unresolved ownership relations (Sopić and Gavrilović 2013).

Another essential factor in the development of Savamala district as a trade center was the construction of the central railway station at the end of the nineteenth century (1881–1885) by Viennese architect Wilhelm Von Flattich. It was one of the

Fig. 5.6 Karadjordjeva Street, Savamala. Photographer Biljana Arandelovic. Rights reserved

Fig. 5.7 Karadjordjeva Street, Savamala. Photographer Biljana Arandelovic. Rights reserved

Fig. 5.8 Residence of Princess Ljubica (Konak Knjeginje Ljubice). Photographer Biljana Arandelovic. Rights reserved

Fig. 5.9 Manak's House (Manakova kuća) on the corner of Kraljevića Marka and Gavrila Principa Streets in Savamala, a cultural monument since 1963. The house was built in around 1830 and is representative of Savamala's early urban development. Photographer Biljana Arandelovic. Rights reserved

most extravagant buildings in the country at the time and became active as the main railway station in 1884 when its first connection was established between Belgrade and Niš (Gajović 2013). The building was declared a cultural monument in 1981, and in 1983 it was declared to be cultural heritage of great importance (Figs. 5.10, 5.11, and 5.12). The building has not served as the central train station since December 2018, and it will be redesigned as a museum of Serbian history as part of the Sava Square reconstruction, where construction works are already in progress. The location in front of the museum is planned as the home for the future highest monument in Belgrade, the controversial Stefan Nemanja sculpture.

To better explain the position of Savamala district within the city, it is important to emphasize that its outer edges became the administrative center of the state in the middle of the nineteenth century, when a variety of ministry and government buildings were built. The buildings and area have kept their functions until today, but the Savamala district itself was left to stagnate, that is, until about a decade ago when its cultural renaissance began.

Construction between the two wars was intense, and new squares and streets appeared. Two bombings during World War II destroyed a large part of the area, and despite its great historical significance, Savamala district was forgotten during the twentieth century, especially during the former Yugoslavian Socialist regime. After the Communists came to power, many of the buildings were nationalized. Also, during the communist period, Savamala was considered to be a legacy of the capitalist era and was turned into a transit route for international traffic through Belgrade. Savamala's Karadjordjeva Street became a major route for heavy transport through the city center. The district entered the twentieth century as a central business and cultural area, and left it as a transit zone.

Its historical heritage was left behind and many buildings stayed abandoned. All of Savamala's heritage, both cultural and architectural, dates back to the period before World War II, when it was boosted as a major trade area and an important communication hub, with bus and train stations in its proximity. Numerous crises have left their mark on Savamala district and many buildings have become seriously rundown. After World War II, Savamala was in a constant state of dilapidation and was marked as a socially disadvantaged and underdeveloped district, at a time of many socio-spatial conflicts. The former Sava port, once the pride of Belgrade during its glory days at the beginning of the twentieth century, was filled with abandoned wrecks for a long period of time. Restitution was a necessary process but it proceeded slower in Serbia than in other former socialist countries. There were many open property issues, which caused further difficulties in solving problems relating to the cultural and architectural protection of important historical buildings. The result was inefficient restitution and restricted architectural maintenance, which directly increased the architectural decay in Savamala district (Roter Blagojević 2014).

The idea of coming down to the river has long been a topic in Belgrade, but for many years it was impossible due to the waste collection area which had already taken up the riverbank. Finally, the city came down to the Sava River in the twenty-first century, but thanks to the small-scale cultural projects and creative

Fig. 5.10 Former central railway station building at Sava Square in Savamala. It is in the process of being redesigned as a museum of Serbian history. Photographer Biljana Arandelovic. Rights reserved

Fig. 5.11 Former central railway station building in Sava Square in Savamala. It is in the process of being redesigned as a museum of Serbian history. Photographer Biljana Arandelovic. Rights reserved

Fig. 5.12 Interior space of the former central railway station building. Photographer Biljana Arandelovic. Rights reserved

clusters that had completely transformed Savamala district into an unplanned cultural zone.

New changes in the political scene in Serbia in 2000 also led to changes in Savamala district, and massive redevelopment projects have been proposed since then. Currently, several megaprojects are being developed, although they are facing much criticism, some of them directly located in Savamala. Thanks to its prime location on the waterfront (Fig. 5.13), Savamala district has often been included in large-scale architectural projects. Three of these flagship projects from Belgrade's Development Strategy, all designed by star architects, are City on Water by Daniel Libeskind and Jan Gehl, the Beko Masterplan by Zaha Hadid Architects, and Beton Hall Cloud by Sou Fujimoto Architects.

Among them, the Belgrade Waterfront megaproject is the most significant and ambitious and will clearly bring radical changes to the area when it is completed (Figs. 5.14 and 5.15). Currently, only this project is under construction and the other projects have no status. The first phase of the Belgrade Waterfront construction started in 2016 (Fig. 5.13).

Megaprojects bring with them insecurity to cities such as Belgrade, because many planned projects in the past were started but have never been completed. Belgrade follows the latest trends in urban development, although it is not always feasible to put them into practice (Arandelovic et al. 2017). The attractive location of Savamala district is in imminent danger of becoming a playground for corrupt city government officials and powerful private investors who hide under the theme of urban development and economic prosperity (Cvetinović et al. 2016).

5.2 Savamala's Architectural and Cultural Heritage in Modern Circumstances

5.2.1 Belgrade Cooperative Building—Geozavod Building

The Belgrade Cooperative Building, located at 48 Karadjordjeva Street in Savamala district, was built as a monumental corner building in the period 1905–1907 in the style of Academism by famous Serbian architects Andra Stevanović and Nikola Nestorović, directly influenced by French architecture from the beginning of the twentieth century (Djurić-Zamolo 2009). It was built as an administrative office building for the Belgrade Cooperative, the first Serbian insurance company, established in 1882, and until World War II it belonged to the joint-stock company of Belgrade cooperatives. The construction of the Belgrade Cooperative was significant for the economy of Belgrade and Serbia. Because in the late nineteenth century illegal lending was a common occurrence, and rates were as high as 50% per year, the aim of the Belgrade Cooperative was to financially support industrial development, trade, manufacture, and crafts. The cooperative actively invested in Belgrade's economy, supported by Belgrade's most successful businessmen of the

Fig. 5.13 Savanova promenade, part of Belgrade Waterfront project. Photographer Biljana Arandelovic. Rights reserved

era. During the First World War, it was relocated to Marseille, but after the war it never adjusted to its new conditions and lost its initial character forever. It terminated its operations in 1946 (Aleksić 2012).

The building was constructed under the patronage of Luka Ćelović (Peacock 2013), an eminent merchant of that period, who was the director and majority shareholder until his death in 1929 (the property was bequeathed to the University of Belgrade). Ćelović financed not only the construction of Belgrade Cooperative, but also Hotel Bristol, two of the most famous architectural symbols in the Savamala district. Belgrade Cooperative was built on the site of the former Little Market, which is considered to be Savamala's center. Furthermore, the building represents the beginning of the formation of modern Belgrade on the Sava coast (Pavlović-Lončarski 2005).

This authentic building is believed to be one of the most architecturally significant buildings in Belgrade, the first one with a representative lavish hall and a monumental staircase leading to the upper floor of the banquet halls, decorated with German sculptures. It was designed in an eclectic spirit, with elements of neo-Baroque and Art Nouveau, richly decorated with carvings on the facades (Aleksić 2012).

It was built using the most modern construction process of that time, with reinforced concrete, used here for the first time in Belgrade. After World War II, it was the headquarters of several institutions, in particular, the Geozavod, after which it was assigned to the Institute of Geophysics, though today it is still known as the

Fig. 5.14 Belgrade Waterfront project in development. Photographer Biljana Arandelovic. Rights reserved

Fig. 5.15 Braće Krsmanović Street in Savamala district with Belgrade Waterfront in the background. Photographer Biljana Arandelovic. Rights reserved

Fig. 5.16 View of the Geozavod Building Street in 2013. Photographer Biljana Arandelovic. Rights reserved

Geozavod building. The building has also served as the Ministry of Justice and the Judicial Academy. None of the institutions that occupied it sufficiently took into account the historical value of the building, which together with the hotel Bristol forms a representative unit (Roter Blagojević 2014).

Belgrade Cooperative building (Figs. 5.16, 5.17, 5.18, 5.19, and 5.20) has been under protection since 1966, and in 1979 it was declared to be cultural heritage of great importance. It is a visual highlight of the new development of Savamala district after being fully restored in 2014. The biggest challenge to its reconstruction was the recovery of the huge dome.

The building is currently occupied by the Dubai firm Eagle Hills, the developer of Belgrade Waterfront. A large model of the Belgrade Waterfront project is on show in the central hall, enabling visitors to see the future vision of the Sava amphitheater after its planned transformation into a residential and business district. Also, an exclusive restaurant opened on the first floor of the building in 2016, Salon 1905, making its lavish interior available to the public.

Fig. 5.17 View of the renovated Geozavod building in 2018. Photographer Biljana Arandelovic. Rights reserved

Fig. 5.18 Geozavod building after reconstruction. Photographer Biljana Arandelovic. Rights reserved

Fig. 5.19 Reconstructed Geozavod building area. Photographer Biljana Arandelovic. Rights reserved

Fig. 5.20 View of the Geozavod building from Kraljevića Marka Street. Photographer Biljana Arandelovic. Rights reserved

5.2.2 Hotel Bristol

Another masterpiece of Belgrade's architecture, located at 50 Karadjordjeva Street in Savamala district, is Hotel Bristol, designed by Nikola Nestorović. The three-storey building of Hotel Bristol was initially built to house insurance and credit cooperative societies between 1910 and 1912. Luka Ćelović, the founder of Belgrade Cooperative Building, together with the brothers Krsmanović, invested in the construction of the hotel, one of the first modern hotels in Belgrade. At the time of its construction, it was one of the biggest buildings in the city.

The external composition of the building is in the style of Academism, with rich art nouveau decoration, in the same style as the Belgrade Cooperative Building, since architect Nestorović was the author of both. Nestorović also achieved perfect harmony between academic concepts and art nouveau decorations in 1907 with Merchant Stamenković's House on Kralja Petra Street—the popular "building with green tiles" (presented in the chapter: The urban, political, and socioeconomic rise and fall of Belgrade through its history). In style, the building represents the mature phase of Belgrade art nouveau (Figs. 5.21 and 5.22).

After the construction of the hotel, the street started to develop in a new direction. Around it, new houses belonging to rich traders began to rise. This was the period in which Savamala district transformed into a modern and well-decorated

Fig. 5.21 Hotel Bristol in Savamala. Photographer Biljana Arandelovic. Rights reserved

Fig. 5.22 Hotel Bristol in Savamala. Photographer Biljana Arandelovic. Rights reserved

neighborhood next to the Sava River. In fact, Savamala was the center of the cultural life of Belgrade in the nineteenth century. Cultural life was linked to hotels and taverns, where balls and theatrical performances were often organized, giving Hotel Bristol an important role to play in the cultural development of Belgrade. The first modern theatrical performances, film screenings, and royal balls were organized in its rooms. Apart from Hotel Moscow, Hotel Bristol was the most modern hotel in Belgrade for a long time, and it was able to accommodate the most eminent guests, royals, and diplomats.

Hotel Bristol was the architectural symbol of the transformation of Belgrade from a backward oriental town into a modern European capital. However, during the periods of crisis that followed, the hotel building deteriorated, along with the state in which it was built.

After World War II, in the new circumstances in which the state usurped private property, Hotel Bristol was given in the 1960s for use by the Ministry of Defense, or to be more precise the Military Institute Dedinje. The hotel continued to work, but in its residential rooms lived the families of military personnel displaced from the territory of the former Yugoslavia.

The new circumstances regarding the current urban redevelopment of Savamala will directly affect the future of the Hotel Bristol building. The location of the hotel is such that it has become part of the large Belgrade Waterfront development project.

According to an agreement between the government and Eagle Hills company (the Belgrade Waterfront investor), the company will renovate and manage a range of facilities in the Sava amphitheater, including Hotel Bristol, the Geozavod building, the main train station building, and a group of buildings in the Savamala district. These objects will remain in state ownership as they are under state protection. It is not completely clear at the moment what will happen with the hotel. Hotel Bristol was closed in August 2018, and the building is now officially part of the Belgrade Waterfront complex. Its precise new role and function after the renovation process will be known soon. Its renovation is not planned in the first phase of the project. However, in this case, as in general with flagships projects in Belgrade, long-term planning is certainly not possible (Arandelovic et al. 2017).

5.2.3 *Vučo's House on the Sava River*

The house belonging to merchant Djordje Vučo, known as Vučina house, on the Sava River, was built in art nouveau style in 1908 by architect Dimitrije Leko at 61 Karadjordjeva Street in Savamala district. Although the basic concept of the building is based on academicism, the main facade, with a unique surface covered with numerous decorative elements, clearly defines it as an art nouveau building (Fig. 5.23). The building has three facades but the central one, facing the Little Market and the Belgrade Cooperative, is dominant. The decorative plastic applied to it is very diverse, ranging from very simple geometric or floral to very pronounced with typical art nouveau representations of female heads.

The two-storey building was designed with a dual purpose, as a shopping zone in the ground-floor area and a residential area above this. Its interior was a novelty for Belgrade's architecture, with a dominant staircase. The building was damaged in World War II, and during its reconstruction in 1946 the elements that made this interior special disappeared.

Vučo's house is the counterpart building to the Belgrade Cooperative and it stands opposite the location of the former Little Market. The building is also a cultural monument and a unique example of one architect designing all of the houses for one investor. This is the first of four houses that architect Dimitrije Leko designed for the Vučo family (Pavlović-Lončarski 2005).

Another novelty regarding this iconic building is a modern bistro called Mali Pijac (Fig. 5.24), named after the historic square where the Little Market (Mali pijac) once stood. It was opened in the ground-floor area in October 2018 according to the design of Belgrade-based Studio Fluid, and it has already become a prominent venue in Savamala district. Mali Pijac bistro covers an area of 250 m^2, and it fully represents the new age in Savamala district that comes parallel with Belgrade Waterfront—for the business elite.

5.3 Cultural Renaissance of Savamala District

The growth of creative communities, combined with an increasing role of artists in shaping city neighborhoods, has become a widespread and well-known trend in Western European cities. But a city development strategy that includes the strengthening of creative activity is a new phenomenon for Belgrade. During the Milosević regime, the degradation of culture was obvious, and the negative balance was enhanced by the international embargo. Nevertheless, the young and progressive element of the population has found a way to nourish contemporary achievements and cultural trends. The transition period beginning after 2000 brought new challenges for the Serbian creative scene. Awareness of how much profit investment in culture can bring is not sufficiently developed in Serbia. Public investment in culture has always been at a low level, while the private sector is not strong enough and not interested enough in supporting larger cultural and creative projects (Jocić 2019).

Belgrade has always been, even in times of crisis, a kind of broader regional center. A decade of crisis has not completely destroyed the reputation that has been created over many years. Terms such as creative industries and the creative economy are a new appearance in Serbia. However, the development of the creative economy in Serbia is connected to a series of activities that date back even further into the past, when Yugoslavia enjoyed a worldwide reputation and Belgrade was its capital. Belgrade has great potential to be a creative city, since it offers a rich historical and cultural heritage. Savamala district stands out in this and it has

Fig. 5.23 Vučo's House, Savamala. Photographer Biljana Arandelovic. Rights reserved

established itself as the first area of Belgrade with a broader and more influential creative scene, which, since 2007, has created an urban phenomenon that has made a real impact.

5.3.1 Magacin Cultural Space and Cultural Center Grad as the Pioneers of the Cultural Renaissance of Savamala District

The cultural renaissance of Savamala district began in 2007 and transformed Savamala into a creative district through various informal research activities, small-scale cultural projects, and creative clusters. The first transformation of this kind began in 2007 with the inauguration of Magacin cultural space (MKM in Kraljevića Marka cultural center). This location now has a new neighborhood, Belgrade Waterfront. Magacin is the former warehouse of a publishing company (the translation of the name Magacin is warehouse), which was reconstructed as a cultural center in 2007, after which it played a crucial role in Savamala's cultural renaissance.

Fig. 5.24 Bistro Mali Pijac. Photographer Biljana Arandelovic. Rights reserved

The year 2007 is significant because it is when the local government released an open call in which five organizations received empty spaces at their disposal. This was legally possible through a mediator between the city and the organizations, Belgrade Youth Center. In this way, the City of Belgrade officially entrusted the space to Belgrade Youth Center, in order to assign it to different organizations. Prior to this, the city had taken on the responsibility for renovating the space. However, this never happened, which consequently led to a situation in which the organizations selected at the open call to use the space did not sign any contract with the city. The result was their illegal status with the silent approval of the authorities. This consequently brought to light different problems such as deterioration of the space and, most importantly, the absence of infrastructure investment.

Despite these administrative problems, in the meantime, MKM positioned itself as a prominent events space for the independent cultural scene in Savamala, offering rooms for independent theaters, workshops, exhibitions, debates, and other cultural events. Unfortunately, Belgrade Youth Center tried to move the organizations out in 2014, with the goal of renting it as commercial space. They were entitled to do this because a contract was never signed. In the end, an association known as the Independent Cultural Scene of Serbia (ICSS) became involved in the problem, as some of its member organizations were using Magacin Cultural center space, and they started a campaign to ensure that the Cultural Center remained in this space. This was not the only case in which this area was exposed to attempts to convert it into a commercial zone. These attempts are constant.

There have also been attempts to make an arrangement between the formal user of the space—Belgrade Youth Center—and the Secretariat of Culture of Belgrade, with the aim of preserving its current purpose to be an open door for all organizations working in the field of informal research activities, small-scale cultural projects, and creative clusters. Unfortunately, the legal status of MKM is still unresolved.

It is important to mention that the Magacin cultural space MKM is still successfully working today, despite all of the administrative difficulties encountered in recent years. Magacin does not have any employees, but rather it is managed by a community of users according to established policies and principles. Communication between its users and customers is coordinated by an appointed coordinator that changes monthly (MKM 2019).

Another crucial part of the cultural renaissance of Savamala is Cultural Center Grad (Fig. 5.25), located at 4 Braće Krsmanović Street in a building of industrial heritage, an old parquet warehouse from 1884, which was used as a warehouse by the Centrotekstil company after World War II. The full name of the center is Grad (City), European Center for Culture and Debate. The center was founded in 2009 by the NGO Belgrade Cultural Front and the Felix Meritis Foundation from Amsterdam under the leadership of Ljudmila Stratimirović and Dejan Ubović. Since its opening, it has hosted various activities such as concerts, pop-up events, debates, exhibitions, performances, workshops, and lectures, and positioned itself as an engine for new cultural development in Savamala district (GRAD 2019).

Fig. 5.25 Cultural Center Grad, Savamala district. Photographer Biljana Arandelovic. Rights reserved

The building contains a gallery, a conference room, and space for seminars and presentations, as well as a hall for concerts and performances. The institution's program continuously attracts the attention of visitors, and this has been the case since it was established. In fact, this positive initiative has outgrown its local potential and has long since become not only a regional but also an international cultural institution. Without doubt, Cultural Center Grad was one of the key factors in Savamala becoming a regional creative district. The center is still entirely active, although it is located near the Belgrade Waterfront construction site. For now, there is no danger of its closure.

5.3.2 Urban Incubator Belgrade Project

The third important project that has influenced the development of Savamala as a creative district, in addition to the opening of Magacin cultural space in 2007 and Cultural Center Grad in 2009, is the Urban Incubator Belgrade project. This project was initiated in 2012 by Belgrade's Goethe-Institute in collaboration with curator Tim Rieniets, also supported by the City of Belgrade and the Municipality of Savski Venac (to which Savamala district belongs). It involved architects, artists, and Savamala residents, with the main objective of supporting the idea of change in Savamala and facilitating contemporary discussion on art and the city and other creative city topics.

The Urban Incubator project has raised some central questions, about what the role of art and culture is in urban development, and whether it works through self-initiated, bottom-up actions or through a guided, top-down process. It also asks the question of how art and culture can be used to improve urban areas. The project opens dialogue about what consequences art and culture can have, and how they contribute to the gentrification process (Krusche and Klaus 2015).

The Urban Incubator Belgrade project embraced more than ten local and international projects by teams from Serbia, Germany, and Switzerland which were selected through competitions within the areas of architecture, community engagement, urban development, and art. The project teams were invited to work in Savamala from March to November 2013. Some of the institutions and authors involved in the project were the University of Technical Sciences (ETH) Zurich, raumlaborberlin, the art collective Third Belgrade (Treći Beograd), Hamburg University of Fine Arts, Zurich University of the Arts, Nexthamburg (Hamburg), GingerEnsemble (CH), Camenzind (Zurich), Goethe-Guerilla, and Maja Popović and Boba Stanić (Belgrade/Amsterdam).

For example, the art collective Third Belgrade carried out the project *We Also Love the Art of Others*. As explained by Third Belgrade members, the project aimed to invite prominent national and international artists representing different schools to work in the context of Savamala district. These artists were selected on the principle of love, according to the theory that we can only love what we know and understand well (urbanincubator.rs).

Further, the curator, Tim Rieniets, asked the Institute for Contemporary Art Research (IFCAR) for its cooperation in the development of an accompanying project to Urban Incubator, which resulted in the creation of Bureau Savamala by Public City (a research unit of the Institute for Contemporary Art Research). The Bureau Savamala project team was made of Jürgen Krusche (Institute for Contemporary Art Research, Zurich University of the Arts), Philipp Klaus (INURA Zurich Institute), and Belgrade-based political activist and scientist Dobrica Veselinović. When it started, the project was focused primarily on the role of art and culture in progress, and what influence they have over it. This focus led to another crucial topic, gentrification.

Bureau Savamala was divided into six research areas: *Urban Incubator projects* (the projects leaders were evaluated through interviews), *Savamala Residents*, *Experts*, *Buildings and Public Space*, *Media,* and *Statistics*. The *Savamala Residents* area of research was divided into four parts: *Interviews* (with Savamala residents and visitors), *Appeal Profile* (a survey in which people were asked to express their feelings about Savamala using a list of adjectives), *Substitutional Photo Inquiry*, and *Nightwatch*. The results of the *Substitutional Photo Inquiry* were presented as a street exhibition called *Like/Dislike Savamala* on July 19, 2013, and were also published in *Kamenzind* magazine. *Nightwatch* was organized on the same day and was an event exploring the nightlife scene in Savamala.

The third research area was *Experts*, which gathered specialists to share their expertise, including the head of the Savski Venac district administration (to which Savamala belongs). The *Buildings and Public Space* research was made up of photo documentation collected from 2013 to 2014, and included facades, sidewalks, parks, streets, and changes related to them. Within the *Media* research, the project team used local daily newspapers and online magazines to gain insight into their perception on the process of change in Savamala. By observing the media perception on Savamala, conclusions were formed and later presented. The final research area in the Bureau Savamala project was *Statistics*. The project team had access to census data from Savski Venac Statistics Bureau and was able to study demographic changes (Müller-Wieferig and Herzen 2013).

Through its subprojects, this project brought a new dimension to the district through its architectural, educational, cultural, and artistic activities. It also supported civic participation and cooperation. Some of *Urban incubator Belgrade's* subprojects have continued to grow and develop, even though the Goethe-Institute officially finished its project in Savamala.

5.3.3 The Spanish House—The Ghost Building

The Spanish House at 4 Braće Krsmanović Street (next to Culture Center Grad) is one of Savamala's many devastated buildings, left aside after failed reconstruction attempts. It was built around 1880 as a corner palace with a basement, ground floor, and first floor, designed in the spirit of academic architecture from the nineteenth

Fig. 5.26 Spanish House, Savamala district. Photographer Biljana Arandelovic. Rights reserved

century. It is not known how and why it got this name, the Spanish house (Fig. 5.26).

According to data from the Documentation Center at the City Institute for the Protection of Cultural Monuments, in the time of Savamala's glory, this building used to be a customs house. It served later as a warehouse and also as a space for the Serbian shipping company. The Museum of River Shipping was located in this facility after World War II. The Spanish House was nationalized in 1960, like many other buildings in this area, but was then left aside during the period of communist Yugoslavia. At the time when it was built, it was one of the most representative buildings in this part of town.

This ruin was transformed into a creative venue and was used for events (performances, concerts, debates, exhibitions, workshops) within the Goethe-Institute's Urban Incubator project during the summer of 2013. Goethe-Institute got permission for the project from the district government. The building has had no roof for many years (Krusche and Klaus 2015).

There have been some attempts to reconstruct it, but without any success. New developments have brought possible new life for this historical location. According to the planning document for the Belgrade Waterfront project, this street, Braće Krsmanović, will become a pedestrian street while the railway will be relocated. Spanish House is also included in this redevelopment plan. The plan is to renovate the house into a cultural institution. Spanish house will serve as a house of architecture, exhibition, and education on various architectural topics. The city will, in cooperation with the Association of Belgrade Architects, announce a competition for the selection of a solution that will be the basis for the renovation of Spanish House. The new facility will include a multipurpose exhibition space, library and lecture hall, and a restaurant.

5.3.4 *Cultural Organization Mikser*

For some, the name Mikser is associated with the largest regional festival of creativity and innovation, and for others with the large venue space of Mikser House, which was located in 46 Karadjordjeva Street (Figs. 5.27 and 5.28), in the immediate vicinity of Belgrade's iconic Cooperative Buildings—Geozavod and Vučo's House. Mikser House unfortunately closed in 2017, although it achieved remarkable success during the 5 years of its existence. Mikser festival was held for the first time in 2009, in the old mill in the central Belgrade district of Dorćol. The festival was organized by an organization of the same name, Mikser House. It was envisioned as a regional festival of the creative industries, with ideas and innovations in line with current global trends. It gave the opportunity for people working in the creative industries to present their work to a wider audience. In 2005, the same

Fig. 5.27 Karadjordjeva Street in Savamala district. Photographer Biljana Arandelovic. Rights reserved

Fig. 5.28 Mikser House in Savamala district (2012–2017). Photographer Biljana Arandelovic. Rights reserved

organization established Belgrade Design Week, the first conference in Southeast Europe dedicated to creative industries.

In 2012, Mikser festival moved to Savamala district, marking the beginning of a 5-year period of unique urban transformation through creative economic initiatives. Always held in the late spring, the Mikser festival transformed the neighborhood into an open stage, with workshops, readings, concerts, video installations, exhibitions, and performances for locals and visitors to enjoy. The idea was to start major social transformations that would help the authenticity and uniqueness of Savamala district. Other projects by Mikser include the *Ghost Project* (Figs. 5.36 and 5.37), an international showcase for young industrial designers aiming to connect industry with emerging talents, the Young Balkan Designers competitions, and the Soundtrack Call competition.

One of the strategic aims of the Mikser creative festival, besides promoting young regional talents, is to help the transformation of Savamala district. There have been numerous multidisciplinary projects held during the festival (workshops, lectures, competitions, performances) connected with the Savamala urban transformation initiative, which has made the Mikser organization one of the most active players in transforming Savamala into a creative district.

Mikser House was a multidisciplinary initiative and social hub, located in a former abandoned steel factory warehouse in historical Karadjordjeva Street, with

Fig. 5.29 Mikser House interior in Savamala district (2012–2017). Photographer Biljana Arandelovic. Rights reserved

the aim of promoting and developing national and regional creative industries (architecture, new media, visual arts, music, media, and design).

The extensive open space of 1000 m^2 was transformed into a modern room and equipped for cultural, educational, and commercial activities that took place throughout the year (Figs. 5.29 and 5.30). It also hosted the first Balkan Design Store, which included the Balkan Gastro Corner, with selected regional products. This authentic concept, the first and the biggest of this kind in the region, aimed to show products, gain attention, exhibit, and sell at the same location. More than 1,500 concerts, festivals, theater performances, exhibitions, debates, and fairs were organized over 5 years (with more than a million visitors) (Mikser House 2019).

The inability to obtain a long-term, stable contract with the private company that owns the building and the continual increase in the monthly rent made the business model of Mikser House in Savamala district completely unsustainable. Mikser House never received any grant from the city or state. Five successful years of Mikser House have proved that the creative sector can be a catalyst for social, economic, and urban development, even in a city like Belgrade, where the state does not participate, or rarely participates, in the development of the creative industries. Having in mind the Serbian context, the absence of local institutional support, the success of Mikser to transform an abandoned warehouse in Savamala district, and its initiative to reactivate the local community through cultural, artistic,

Fig. 5.30 Mikser House interior in Savamala district (2012–2017). Photographer Biljana Arandelovic. Rights reserved

and educational activities, it can be concluded that it achieved fascinating results. The Mikser concept of a cultural and social center as a catalyst for urban regeneration has become a phenomenon. This is evidenced by the numerous reports by leading world media on Mikser House, its festivals, and all of its cultural activities. The Guardian, CNN, Mashable, Lonely Planet, Business Insider, and Shanghai Daily have all written about Savamala district. In the building of the former Mixer House today are two night clubs, the Hype and the Martinez, both opened in 2017, the year when Mixer House moved out.

The 10th Mikser Festival was held in late May in 2019 in the central Belgrade district of Dorćol, after a 2-year break. It was an open stage of workshops, readings, concerts, video installations, exhibitions, and performances but dedicated to the theme of sustainable development as a new culture of life and work. The Museum of Science and Technology at Dobračina 51 in Dorćol district has been chosen for the central festival location. More than 500 participants took part in nine program areas, including the Shift international conference and the Circulation workshop, an innovations fair, Hackathon for Change, a design zone, CirculArt, Good Food, Eco Bazaar, mixer kids, and an accompanying musical program. The Young Balkan Designers platform was again part of the festival.

5.3.5 Bohemian District

Belgrade's Savamala district has transformed over the last few years into an energetic creative urban district and has become one of the most vibrant Southeastern European urban centers (Figs. 5.33, 5.34, and 5.35). Savamala district has established itself as an attractive alternative scene in recent years. This has consequently started attracting many tourists, most of them young, due to its hedonistic alternative culture and club scene reputation. Savamala has gained a lot of publicity over a relatively short period of time. The British Guardian published a report describing the level of bustle and excitement that one will never find in other parts of the city, pointing out that Savamala district is promoting a new wave of cultural movements and activities. It concludes that creative entrepreneurs based in Savamala are birthing a new image of the city (Coldwell 2015).

There are four main clusters of nightlife in Belgrade, differentiated by the type of music they play. The first is Skadarlija, located in the center of Belgrade with "starogradska" music, which is urban folk music from the early twentieth century.

The second cluster is Upper Dorćol, with Strahinjića Bana Street as its center. The area is also known as Silicon Valley, named after the style choices of the women who go out there (implanted silicone), and it is visited by the new elite that have emerged during the transition period. The music that one can hear in Silicon Valley café-bars is mostly pop. The third cluster is by the Sava and Danube, on the boat cafés known as splavovi (rafts). This is a very important part of the city nightlife as these objects are open through the whole night, and they play a specific type of music called "turbo-folk," which is popular contemporary Balkan music, containing a blend with oriental sound (Medić 2016).

Savamala district is, according to Medić (2016), the fourth and most recently formed cluster. It cannot be identified by only one specific genre of music, as in the previous three cases. Here, there is a mix of genres that have one thing in common; they offer an alternative to what is offered in the other places. Music played and performed in Savamala district is very diverse, including jazz, electronic, alternative, and indie rock.

In Savamala district, it is possible to find new exclusive bars as well as other more informal eclectic venues. Today, it is a hip urban district with a high concentration of young bohemian visitors, a rich subcultural vibrant area with one of the highest bar and club densities in Belgrade. There's a mix of trash and graffiti (Figs. 5.31, 5.32, 5.33, 5.34, 5.35, 5.36, 5.37, and 5.38), pop-up stores, bars and cafes on one side of Savamala, and Belgrade Waterfront building site on the other side.

The construction and demolition of some locations in Savamala, due to the construction of Belgrade Waterfront, has caused the closure and departure of some clubs and bars that have been working there for the last decade. An example is the club Dvorištance, whose location was ruined in 2015 due to Belgrade Waterfront's construction. The owners of Dvorištance have, however, found a new location for the club, which has gained a bohemian reputation since it moved to the old city

Fig. 5.31 Mural of iconic Serbian actor Bata Stojković by Zagreb-based street artist Boris Bare. Photographer Biljana Arandelovic. Rights reserved

center. On the other hand, new bars and clubs have opened in Savamala. There has been a shift, a change, which is happening slowly, in the structure of the people coming to Savamala to enjoy the nightlife. This process has only recently begun, and it is therefore still too early to talk about any final results.

5.3.6 *Small-Scale Cultural Practices, Creative Clusters, and Initiatives with Belgrade Waterfront in the Background: 12HUB, Nova Iskra, and the Do Not Drown Belgrade Initiative*

The social circumstances in Savamala district have gradually generated a number of cultural practices: small-scale practices, creative clusters, and various initiatives (Simeunčević-Radulović and Stupar 2016). The very first activity of this kind was Magacin cultural space in Kraljevića Marka Street, which opened in 2007. This continued with Cultural Center Grad in 2009 and with the founding of Mixer House in 2012. In the same year, the gallery 12HUB was opened in Karadjordjeva Street just opposite Mikser House. This was a creative hub of art and technology with the main aim of promoting digital and performance art, in order to ensure the availability of digital technology research and production for both artists and scientists,

Fig. 5.32 Travnička Street, Savamala. Photographer Biljana Arandelovic. Rights reserved

Fig. 5.33 Braće Krsmanović Street, Savamala. Photographer Biljana Arandelovic. Rights reserved

Fig. 5.34 Braće Krsmanović Street, Savamala. Photographer Biljana Arandelovic. Rights reserved

Fig. 5.35 Braće Krsmanović Street, Savamala. Photographer Biljana Arandelovic. Rights reserved

Fig. 5.36 Ghost People in Savamala. Photographer Biljana Arandelovic. Rights reserved

as well to encourage their mutual cooperation. Gallery 12HUB left the space it occupied in Karadjordjeva Street in 2016 at the request of its owner, the Ministry of Defense, but continued its work through the online promotion of artists, international projects, and educational programs, and as a visiting program in various partner venues in Belgrade. This process can be seen in the context of the impact of the Belgrade Waterfront project on the alternative cultural scene in Savamala district. This space successfully worked as a nonprofit organization for 3 full years and was dedicated exclusively to the development of contemporary art practice and the promotion of young artists. This demonstrated that the government did not recognize the importance and value of investments in culture and the development of the art scene in Savamala district.

In the same year, 2012, a creative hub called Nova Iskra started work, an independent and innovative platform for the professional development and affirmation of young designers and creative workers, the first of its kind. It was founded with the objective to interlink creative potential with industry, trade, and the service sector in Serbia and to offer working space and support for the professional development of young creative people. When it began in 2012, the Designers Lab within Nova Iskra, for example, provided working space for 18 young designers in Savamala. In addition, this creative hub has organized educational and mentoring programs, creative camps, webinars, and multidisciplinary and experimental programs, as well as designer in residence programs.

Fig. 5.37 Ghost People in Savamala. Photographer Biljana Arandelovic. Rights reserved

Nova Iskra developed from Savamala district, Gavrila Principa Street, and soon evolved and spread to other parts of the city to become the most ambitious culture hub and coworking space in Belgrade. Nova Iskra Zemun was opened in 2018 in the municipality of Zemun, and most recently, a modern Nova Iskra Workspace has opened in 2019 in Dorćol. The newest location in Dorćol is much bigger than its Savamala location, with event space for up to 150 guests, and it is able to host a growing community of Belgrade start-ups, freelancers, and creative entrepreneurs. The joint working environment at Nova Iskra Workspace extends to over 1,200 m^2, and it is divided into several functional zones and several floors. Nova Iskra is the most successful creative hub that has developed from Savamala, completely corresponding to global trends in the working culture of the creative industries (Nova Iskra 2019).

The Ne da(vi)mo Beograd (Do not drown Belgrade) Initiative is the best example of an initiative developed in Savamala district as a response to the Belgrade Waterfront project. It aims to bring together organizations and individuals who are interested in urban and cultural policy, as well as in the sustainable development of the city, in order to demand the fair use of common resources. Most importantly, the initiative encourages the direct involvement of citizens in the development of their environment.

Fig. 5.38 The mural *Waiting for the Sun* in Savamala district, inspired by a vintage Hungarian matchbox from the post-war Socialist era, was made by Aleksandar Maćašev for the occasion of the Mikser Festival in 2013 on the initiative of the Mikser organization and Ceresit Facade Paints. The mural is located on the wall of the house just opposite Mikser House in Karadjordjeva Street. Photographer Biljana Arandelovic. Rights reserved

They gather around a common goal: to stop the degradation and looting of Belgrade on behalf of megalomaniacal urban and architectural projects, primarily "Belgrade on water" (NDMBGD 2019). The initiative, whose trademark has become a big yellow duck, was formed in 2015, and since then it has organized a number of activities and protests in order to keep up the criticism of the Belgrade Waterfront project. Public attention was attracted by the protest held on April 26, 2015, in front of the Belgrade Cooperative building, during the signing ceremony of the Belgrade Waterfront project. The public struggle against this megaproject and the resulting change in Savamala's character as a creative district has provoked further initiatives of this kind. Resistance to this change in Savamala's urban character, particularly in response to the Belgrade Waterfront project, was inevitable. The intensity of activism has grown in parallel with the new construction sites, and controversy has followed the procedures that relate to this new urban development (Cvetinović et al. 2016).

During 2016 and 2017, the Do not drown Belgrade initiative organized eight mass street protests, and in 2018 it developed into a political initiative that took part in Belgrade's city elections.

Since the Belgrade Waterfront project was announced in 2014, changes in Savamala have intensified. Accordingly, with the development of this megaproject, many new high-end bars have opened. What had started as an assembly of alternative culture venues has slowly begun to resemble Strahinjića Bana Street from Upper Dorćol, the so-called Silicon valley. Savamala district is in the process of deep change: on the one hand, it has the Belgrade Waterfront building site, and on the other hand, culture venues which are the result of a long process of creative and culture-led urban development. Savamala's development into a creative district was completely unplanned, but there is a necessity for a clear strategic vision of how these two can coexist. At this point, it seems that the future development of Savamala is not heading in the direction of being a creative district. Some important institutions have left the area (Mikser, 12HUB) and not many as new and as big and innovative have appeared. We are witnessing change but can't predict a possible outcome.

5.4 Beton Hall Complex

The Beton Hall complex located at 2–4 Karadjordjeva Street on Savamala promenade is a reconstructed location that hosts several trendy restaurants and clubs. This 300-m-long complex is located directly on the Sava riverfront and is one of the first reconstructed spaces in the area. Belgrade Waterfront project is in its vicinity. Beton Hall was constructed in the late 30s (from 1936 to 1939) on the right bank of the Sava, as a customs warehouse. It was not planned for commercial use; in fact, the development of the complex as we know it today happened spontaneously. Initially, there were five cranes, only one of which has been preserved. This yellow crane, the only one remaining, is considered a cultural monument and is set at the beginning of the complex as a unique portal through which pedestrians and cyclists pass. The complex is part of the industrial architectural heritage.

Over time, the warehouses ceased to be used for storage purposes and this site became a garbage lot. All 17 warehouses within the complex were then rented as storage from 1993 until 2005. The Beton Hall complex owner, the City of Belgrade, wanted to rearrange this attractive location, which was done gradually over a period of years. An architect, Aleksandar Rodić, was the first to rent several spatial units in 2005 and turn them into a coffee club, art gallery, and furniture salon. His example was soon followed by other investors, who opened several cafes, restaurants, and offices.

The potential of the location was obvious and very soon, in 2011, the city government announced a major international open competition for a project to connect the hall and the fortress. The problem of connecting Kalemegdan Fortress and the river had existed for a long time, and the reason for this lay in the fact that the access to the river was cut off because of the construction of the railway in the period after World War II. The winning Beton Hall project, a pedestrian square made of suspended ramps and designed as an access point from the riverfront to the

Fig. 5.39 Beton Hall, Savamala. Photographer Biljana Arandelovic. Rights reserved

historic core, was created by Sou Fujimoto Architects. This was another unrealized flagship project in Belgrade. Due to political changes in the city government and the excessive costs of implementation, the whole project has been postponed.

Because of the popularity of Beton Hall (Figs. 5.39, 5.40, 5.41, and 5.42), the new city government has made the decision that the connection between Kalemegdan and the river is a priority. This has led to a new project that is currently in the process of realization. This project is much smaller than the initial vision from the 2011 competition and is actually the artwork of Richard Deacon and Mrdjan Bajić, consisting of a new pedestrian connection, with public artwork situated directly at Beton Hall (Fig. 5.43).

The Beton Hall complex is one of the busiest leisure locations in Belgrade's center, with a constant flow of people coming by the river for that special atmosphere. The high concentration of restaurants affords a hedonistic and energetic alternative to the city crowd. The complex is positioned in a very attractive location, in the center of Belgrade at the foot of Kalemegdan and the confluence of the Sava and Danube, with a view toward New Belgrade and the Museum of Contemporary art. The long promenade, including cycling paths, connects the Beton Hall complex and Belgrade Waterfront.

Fig. 5.40 Beton Hall, Savamala. Photographer Biljana Arandelovic. Rights reserved

Fig. 5.41 Beton Hall, Savamala. Photographer Biljana Arandelovic. Rights reserved

Fig. 5.42 Beton Hall, Savamala. Photographer Biljana Arandelovic. Rights reserved

Fig. 5.43 A new pedestrian connection between Kalemegdan Fortress and the river. Photographer Biljana Arandelovic. Rights reserved

5.5 Reimaging Belgrade Through Savamala

The path of Savamala district's development dates back two centuries. This place has witnessed both prosperous and poor epochs. Savamala boasts a rich social, cultural, and historical heritage, which gives it even greater importance and specificity. Since its early days, political involvement has been of crucial relevance for the development of Savamala district, both in prosperous times and periods of decline. The best example for this is the period after WWII when Savamala was left in a state of deterioration after the political decision that all attention should be directed toward New Belgrade's development. Today again, politics will decide on the future of Savamala district, because it is political will that will determine whether to stop the development of Savamala as creative district or allow what was started in 2007 with Magacin cultural space initiative and Cultural center Grad to continue.

Time will tell whether Savamala will lose the character of a creative district, which it has acquired over the last decade of activities that have attracted regional and international attention, or whether it will fail to comply with its newly acquired character brought on by the Belgrade Waterfront megaproject.

Savamala district is one of the most important topics of interest in Belgrade today. The creative district of Savamala arose spontaneously starting from 2007, and it is still changing. It is evident that the creative industry has played an immense role in its transformation. For the first time since the beginning of the twentieth century, Belgrade's citizens are visiting the area more than ever, and Savamala district has gained international attention and visibility. During the last decade, it has established itself as an attractive creative scene with small-scale art and creative clusters that do not belong to institutionalized organizations. Nevertheless, Savamala still looks wild and alternative, with its neglected courtyards and attics, abandoned basements, warehouses, and various historical palatial spaces. Deteriorated spaces are visible all around.

However, the Belgrade Waterfront project building site occupies an immense surface. It is not completely clear for some plots whether they are a building site or neglected building, or where the border is between them. Many property issues are still unresolved and future legal procedures are very complicated. Compared with these problems, art and culture initiatives look utopian. During the biggest crises in modern Serbian in the 1990s, a large number of properties owned by the municipality vanished from the records and were reassigned through manipulative actions to private owners. The regulation of legal and ownership issues in Savamala district is still a current topic. After returning property to the descendants of its pre-war owners, it was known exactly what would be available for use and which property was available to be used for public purposes, including use by creative organizations. It was at the beginning of the twenty-first century that the municipality reclaimed the public facilities in Savamala district and became their new owner, thus making it possible to offer them to creative practices at reasonable prices (Jocić 2019).

The city authorities have always had big plans and great stories to tell, but the biggest change in Savamala district so far has come from cultural initiatives led by individual enthusiasts. Inclusion of the Republic and the local government was not crucial for the creative development and regeneration of Savamala district; moreover, it was kept to a minimum. Apart from the Ministry of Culture, which helped only a few initiatives with very limited resources, there were no other significant contributions (Jocić 2019). The cultural initiatives from Goethe-Institute Belgrade, Magacin and Grad culture spaces, as well as Mikser festivals and events, were important steps forward in the process of urban transformation in Savamala and its cultural renaissance—making it visible on an international scale. The development path of Savamala as a creative district clearly shows that the success of the first creative organizations (Grad, Magacin, Mikser) at the international level shaped the way forward for other organizations.

Belgrade cannot compete with big cities like Berlin, the most famous example of a city that profiled itself as a world creative city. Clear examples of transformation in Berlin are the transformations of Kreuzberg in the West and Prenzlauer Berg in the East just after the fall of the Wall into nonsubsidized arts scenes (Arandelovic and Bogunovich 2014). The concept of a creative city began to be used for the first time in Berlin in 2000 for marketing purposes when the local government started an intensive promotion of Berlin as a creative city through diverse marketing campaigns. Berlin was then promoted as a place for creative industries and new business start-ups, in order to encourage the arrival of creative companies and individuals (Arandelovic 2018).

The best illustration of cultural renaissance is Neukoelln, one of Berlin's most problematic districts, which has been transformed in the last couple of years from a dangerous problematic and socially disadvantaged area into a roaring bohemian district where new bars, galleries, and studios are opening every week. Berlin's social trouble spot, where immigrants clashed in the streets, has now become a synonym for positive urban change in which the mix of influences and diversity presented here creates new forms of design.

Neither can Belgrade compete with some cities in more developed countries such as Graz, which became European Culture Capital in 2003, resulting in visible changes in the city through new landmarks, the Murinsel and Kunsthaus, both designed for the occasion of European Culture Capital. Graz applied to the UNESCO Creative Cities Network, finally gaining the Design City title in March 2013. This all enriched the image of the city through both large- and small-scale cultural projects and creative clusters (Arandelovic 2015).

Belgrade doesn't have the economy or infrastructure to boast a creative sector on the same scale as Berlin or Graz, but Belgrade definitely has the potential to become a regional leader in this field. Savamala district has space available for new businesses, and its waterfront space and good transportation infrastructure are also beneficial for its future development. Cultural districts like Savamala, as mixed-use areas, might help change the image of Belgrade. City authorities should not miss this opportunity to develop this historical district in the right direction and to integrate it better within a wider urban structure. The high-profile flagship project of Belgrade Waterfront should not destroy what was developed here over the last decade, which is pure treasure through diverse informal art activities, small-scale cultural projects, and creative clusters. The cultural renaissance of Savamala district can bring much more. This was the first step, the hardest, and most important one.

References

Aleksić V (2012) Beogradska Zadruga i njena uloga u privrednom razvoju Srbije krajem XIX i početkom XX veka. [Belgrade Cooperative and its role in the economic development of Serbia from the late 19th to the early 20th century] Bankarstvo 6: 108–133

Arandelovic B (2015) Graz, UNESCO city of design and historical heritage. Cities 43:78–91

Arandelovic B (2018) Creative city Berlin. In: Public art and urban memorials in Berlin. Springer International Publishing, pp 1–26

Arandelovic B, Bogunovich D (2014) City profile: Berlin. Cities 37:1–26

Arandelovic B et al (2017) Belgrade: imaging the future and creating a European metropolis. Cities 63:1–19

Coldwell W (2015) Belgrade's Savamala district: Serbia's new creative hub. The Guardian February 7 2015

Cvetinovic M, Maricic T, Bolay J (2016) Participatory urban transformations in Savamala, Belgrade—capacities and limitations. Spatium 36:15–23

Djurić-Zamolo D (1977) Beograd kao orijentalna Varos pod Turcima 1521–1867 [Belgrade as an oriental town under the Turks from 1521–1867]. Muzej Grada Beograda, Belgrade
Djurić-Zamolo D (2009) Graditelji Beograda: 1815–1914. Muzej grada Beograda
Gajović N (2013) Paris—Venice—Savamala—Istanbul. Camenzind 4:62–67
GRAD (2019) O Gradu, Evropski Centar za Kulturu i Debatu GRAD. http://www.gradbeograd.eu/info.php. Accessed 20 Jan 2019
Jocić N (2019) Creative economy in a post-socialist transitional context: spatial distribution of creative activities and their interrelation with urban milieus in city quarters of Belgrade. University of Bamberg Press, p 236
Krusche J, Klaus P (eds) (2015) Bureau Savamala Belgrade: urban research and practice in a fast-changing neighborhood. Jovis, Berlin
Medić I (2016) The soundscape of change: the reculturalization of Savamala. Musicol Annu 52 (2):39–53
Meriage LP (1978) The first Serbian uprising (1804–1813) and the nineteenth-century origins of the Eastern Question. Slavic Rev 37(3):421–439
Mikser House (2019) http://house.mikser.rs/en/
MKM (2019) Culture Center Magacin. https://kcmagacin.org/
Müller-Wieferig M, Herzen E (2013) Urban incubator Belgrade: report of programme year 2013. Goethe Institute, Belgrade
NDMBGD Ne da(vi)mo Beograd (Do not drown Belgrade) initiative (2019) nedavimobeograd.rs
Nova Iskra creative hub (2019) novaiskra.rs
Pavlović M (2013) Knez Milos & His 'Litlle Sava Girl'. Camenzind 3:33–36
Pavlović-Lončarski V (2005) Mali pijac na Savi krajem XIX i početkom XX veka [Little Market on the Sava in late 19th and early 20th century]. Nasledje 6:07–118
Peacock L (2013) Remembering Luka Celovic. Camenzind 1:21–25
Petrović J (2010) Istorijski, društveni i ekonomski razvoj Karadjordjeve ulice u Beogradu do 1914 [Historical, social and economic development of Karadjordjeva Street in Belgrade until 1914]. Godišnjak grada Beograda = Annuaire de la ville de Belgrade 135–149
Popović M (1997) Stari Beograd - segment Savske varoši - lociran u zoni planske razrade projekta - Beograd na Savi. [Old Belgrade—segment Sava town—located in the area of the planned development of a project—Belgrade on the Sava]. Nasledje 1:33–37
Roter Blagojević M (2014) The impact of the restitution on the preservation of cultural and architectural heritage and the urban development of modern Belgrade. Limes Plus 11(2):45–56
Simeunčević-Radulović S, Stupar A (2016) Uloga kreativnog klastera Savamale u repozicioniranju Beograda u globalnim hijerarhijama [The role of creative clusters in the process of global repositioning: the case of Belgrade]. Arhitektura i urbanizam 43:29–34
Sopić S, Gavrilović M (2013) Face-lifting facades. Camenzind 1:9–13
Urban Incubator Belgrade. urbanincubator.rs

Chapter 6
Public Art in Belgrade

Biljana Arandelovic

Abstract The Public Art in Belgrade chapter presents the sculptural masterpieces displayed in Belgrade, which are Belgrade's existing landmarks, and it explains three projects, which have not yet been fully completed. The chapter also features the Belgrade public artworks of Ivan Meštrović and the creation of three of Belgrade's visual symbols: Pobednik (Victor), the Monument of Gratitude to France, and the Monument to the Unknown Hero. Besides this, it presents part of the public artwork collection in front of the Museum of Contemporary Art and the monumental Horses at Play bronze composition in front of the House of the National Assembly of Serbia. Three projects that have not yet been fully implemented are presented in the second part of the chapter: The Kalemegdan Bridge project: The Odande dovde (From there to here) sculpture and pedestrian bridge, which goes from Belgrade Fortress to the banks of the Sava; the sound installation and arrow sculpture for the Zoran Djindjić memorial (the first democratically elected prime minister, assassinated in March 2003) and the most controversial, the Stefan Nemanja monument, a traditional figurative representation of one of the most important historical figures for the Serbian people. Murals and street art also have a tradition in Belgrade, and they have gained more attention in recent years with the development of Savamala as a creative district. As part of the public artwork scene in Belgrade, they are also presented in this chapter.

Keywords Public art · Belgrade

6.1 The Position of Public Art in Belgrade in the Twenty-First Century: Introduction

Belgrade is now developing in the right direction, and it has received international attention and publicity in the world media. The BBC and the Guardian are among those who have written about the cultural renaissance of Savamala district and its new growing design scene. This primarily refers to the alternative art scene that has spontaneously developed since 2009 in Savamala district, thanks to the activities of

© Springer Nature Switzerland AG 2020

B. Arandelovic and M. Vukmirovic, *Belgrade*, The Urban Book Series,
https://doi.org/10.1007/978-3-030-35070-3_6

a number of non-institutional cultural organizations and their leaders (Mikser, Magacin cultural space, Grad). The number of tourists is increasing, and the problems that have accumulated for years are being discussed and slowly resolved, though this may take many years (Belgrade Waterfront, the construction of an underground system).

In 2000, a decision was made on the erection and maintenance of monuments and works of sculpture in the city of Belgrade (published in the Official Journal of the City of Belgrade, no. 3 of 17 March 2000), which defines what constitutes a monumental work and how public spaces are defined, as well as the necessary procedures regarding their choice and placement, the necessary approvals and permits, the source of financing, and the duties of maintenance and control. Belgrade's Department of Culture has launched a project to encourage creativity and projects that will contribute to the beautification of some city spaces.

For this purpose, in 2008 a committee composed of nine artists, art historians, urban planners, and architects was established with the aim to set up artworks in Belgrade's public space and to estimate the artistic value of any proposed projects and artworks (including estimating the value of donated sculptural works). The objective was to identify the appropriate urban spaces in Belgrade in which sculpture would be permanently installed and for that purpose to select the works of 10 Serbian artists. The plan was to select artists based on a survey. Invitations to participate in the survey were delivered to 107 artists, art historians, and journalists, 48 of whom responded and suggested 10 contemporary artists whose work they considered deserved to be in public space in Belgrade.

The five highest ranked artists, according to the survey, were invited to set up their sculptures in public spaces, these being Mrdjan Bajić, Kosta Bogdanović, Olga Jančić, and Zdravko Joksimović, while the committee additionally selected five more (Milija Nešić, Vladimir Perić, Dragoljub Raša Teodosijević, Zoran Todorovic, and Dobrivoje Bata Krgović). It was planned that five more artists would gain a chance, through a public call. The idea was to split public art equally, both in the central city area and in the new settlements on the outskirts of the city.

The selected locations were Knez Mihailova Street between Albania Palace and Sremska Street (above the underground passage), the space in front of Belgrade City Library in Knez Mihailova Street, the green area in front of the Madlenianum building in Zemun, in front of Bitef Theatre, the square at the corner of Cara Dušana and French Streets, the underground entrance across from Vuk's monument, the green area in Kamendin, the green area in front of the Yugoslav Drama Theatre, in front of the General Post Office, the space in front of the Faculty of Law, the area around the fountain at the main entrance of the Sava center and a green area in front of its rear entrance, the square in Palmotićeva Street, and the green stretch in front of the promenade along the Danube in Dorćol near Nebojša Tower.

Unfortunately, none of this has actually been implemented in Belgrade yet. Financial reasons are usually cited as the main problem, but there are also problems at the organizational level and in the long administrative process. However, more important reasons for the failure of these projects are political and include interference in the decision-making process by unqualified people from political parties (Vukotić Lazar and Danilović Hristić 2012).

The following text presents the sculptural masterpieces displayed in Belgrade, which are Belgrade's existing landmarks, and it explains three projects, which have not yet been fully completed. It begins with the Belgrade public artworks of Ivan Meštrović: Pobednik (Victor), the Monument of Gratitude to France, and the Monument to the Unknown Hero. They were selected for analysis due to their acclaimed artistic value which has made them visual symbols of Belgrade. Meštrović's works are followed by a presentation of the public artwork collection in front of the Museum of Contemporary Art in Belgrade. The monumental Horses at Play bronze composition in front of the House of the National Assembly of Serbia by Toma Rosandić is featured in the text as a masterpiece of art and one of the most iconic artworks in public space in Belgrade.

The Museum of Contemporary Art possesses a significant collection of Yugoslav art from the twentieth century and displays some of the sculptural masterpieces by prominent authors in front of it. Although the collection is significant and its location is perfect, this sculpture park is not known widely by the general public.

Three projects that have not yet been fully implemented are featured in the second part of the chapter. The Kalemegdan Bridge project is a collaborative work between two renowned artists, British sculptor Richard Deacon and Mrdjan Bajić, a contemporary Serbian sculptor and professor. It is made up of the monumental sculpture From there to here and the pedestrian bridge that goes from Belgrade Fortress to the banks of the Sava. The second public artwork is a planned memorial for Zoran Djindjić, the first democratically elected prime minister, assassinated in March 2003. The last artwork presented is the most controversial, the Stefan Nemanja monument, a traditional figurative representation of one of the most important historical figures for the Serbian people.

Murals and street art also have a tradition in Belgrade, and they have gained more attention in recent years with the development of Savamala as a creative district. As part of the public artwork scene in Belgrade, they are also presented in this chapter.

6.2 The Belgrade Public Artworks of Ivan Meštrović and the Creation of Belgrade's Visual Symbols

Controversy followed the placement of almost all monuments by Ivan Meštrović in Belgrade, the author whose works marked the beginning of modern sculpture in Yugoslavia. Croatian-born Ivan Meštrović was devoted to the idea of the unity of South Slavic nations from his youth. As a student at the Academy of Fine Arts in Vienna, it was recommended that he make a bust of King Petar I Karadjordjević, and this acquaintance with the King introduced him to political circles in Belgrade. It is known that he was interested in the idea of forming a union of South Slavic peoples.

At the International Art Exhibition in Rome, the Croatian and Serbian artists exhibited together in the same separate pavilion, on his initiative. The Pavilion of the Kingdom of Serbia was designed by architect Petar Bajalović and it is where Meštrović revealed his artworks from the Vidovdan Temple cycle. Meštrović came to the idea in 1908, and the first presentation of sculptures was in the Pavilion of the Vienna Secession (Stojanovic and Stanojev 2016).

"*The Vidovdan Temple represents a valuable historical source for understanding the Yugoslav project and it enriches our understanding of the historical enterprise of Yugoslavism. In the absence of any substantial cultural framework for imagining a common South Slavic identity with the same kind and the same power of cultural dissemination in the 1910s, this particular artwork served as a sort of Yugoslav reference point*" (Ignjatović 2014: 833). The Vidovdan Temple cycle comprised various figures complemented with Caryatides, directly inspired by epic traditional ballads about the Serbian Battle of Kosovo. Thus, the name of the temple installation is Vidovdan (St. Vitus's Day), which is the day on which the battle was fought, June 28, 1389.

According to Ignjatović (2014: 854): "*Instead of glorifying and redeeming fallen (Serbian) medieval heroes, the Vidovdan Temple was interpreted as evoking a South Slav Golden Age that existed in the remote past, long before the time of the Great Schism, the split identities of Serbs, Croats, and Slovenes, and the Battle of Kosovo. This was a crucial meaning of the Vidovdan Temple—its architecture, sculpture, and iconography.*" After the exhibition, his works were brought to Belgrade and the state bought them later in 1925. They are now in the National Museum of Serbia.

6.2.1 Pobednik (Victor) by Ivan Meštrović

Some of Ivan Meštrović's work is today in Belgrade in the Museum of Contemporary Art, while some artworks are in Belgrade's public space and are an integral part of the city's visual identity. His work Pobednik (Victor) (Figs. 6.1 and 6.2) is one of the most recognizable landmarks in Belgrade. His Angel of Death sculpture (Fig. 6.3), created as part of the Vidovdan cycle and exhibited for the first time at the International Art Exhibition in Rome in 2011, was erected in Kalemegdan Park in 2011. Designed in the form of a standing female figure with wings, holding a phoenix, the figure was created as the final artistic motif in the pavilion dome at the Rome Exhibition as part of the Vidovdan Temple cycle (Pavlović 2016).

Ivan Meštrović is also connected with the urban development of one of the central city squares, Terazije Square. The city government was determined to offer him, without any competition, a project to make the Terazije fountain. Terazije fountain was conceived as an integral part of a major reorganization of Terazije (a development from 1911). The fountain was planned as a monument to dominate the square and to glorify the victory of Serbia in the Balkan wars, after five centuries of

Fig. 6.1 Pobednik (Victor) by Ivan Meštrović at Kalemegdan Fortress. Photographer Biljana Arandelovic

Fig. 6.2 Pobednik (Victor) by Ivan Meštrović at Kalemegdan Fortress. Photographer Biljana Arandelovic. Rights reserved

Fig. 6.3 Angel of Death sculpture by Ivan Meštrović. Photographer Biljana Arandelovic. Rights reserved

Ottoman rule. It was envisaged to stand in front of the new modern Moscow as a monument of both architectural and sculptural compositions in the spirit of secession.

The planned base of the fountain consisted of a circular pedestal on which there were four figures of seated lions. A 5-m-tall Doric column rose from the center of the pool, which was the base for the bronze figure of the Victor (Pobednik) monument. The column had five-annular thickening, which symbolically represented five centuries of slavery under the Ottoman Empire. Seventy masks cast in bronze formed the fountain decoration while a 21-m-long relief was planned around the pedestal, in order to complete this rich urban composition. The Victor monument was in the form of a colossal standing figure of a naked man with a falcon in his left hand and a sword in his right. This monument fountain was planned to be made of Bavarian red granite, except for the decorative masks and Victor monument that would have been cast in bronze. Due to the outbreak of the Second Balkan War, the realization of this luxurious fountain project was interrupted.

The statue was placed in storage in the Senjak district. During World War I, the plaster model of the fountain with all of its parts was destroyed. The Victor monument remained intact, thanks to a hidden warehouse where it was left before the war. Continuation of this luxury project after the war was impossible due to its excessive costs. There was also another issue. Erecting a monument to Serbian heroes from the Balkan wars in a new multinational country was no longer a priority. Aware of the uncertainty of the complete implementation due to the lack of funds, Meštrović asked the city government in the mid-1920s to take a stand on the question of the future of the project. The Victor statue was just partially finished at that moment. As the problem was not solved, he filed a lawsuit in 1923 against the city of Belgrade. Soon after, the situation was resolved, and Belgrade municipality became the owner of the Victor statue and the lion heads (fountain parts), with the agreement that the fountain project would be carried out as soon as the financial situation allowed (Borić 2005).

Given the limited financial capabilities, the poured Victor statue was placed on a stone pedestal in 1927 in Terazije Square. However, the bronze sculpture of a naked man, holding a sword in his right hand and falcon in his left, launched a sharp controversy in Belgrade, indeed in the whole state. The daily press analyzed its moral and aesthetic value in detail. Another problem was the fact that this statue was going to have the most prominent position in the city center, in addition to the size ratio of the monument. The avant-garde understanding of public monuments did not fit the conservative ideas regarding their position in urban space. This monument was created in another time, in pre-war Serbia under different social conditions and historical circumstances. Two decades after its original creation the situation had changed.

Thus, the decision was made to place the monument in the western corner of the upper town of Kalemegdan Fortress in 1928. The artist wasn't consulted regarding the location. The figure was placed to be frontally oriented toward the confluence and toward an empty field, which today is New Belgrade.

This was a brilliant move, which gave Belgrade one of its most admirable visual symbols. Here, we can find some parallels with today's situation regarding the raising of the new monumental sculpture of Stefan Nemanja in Savamala's Sava Square, in front of the historical railway station building. The size ratio between the Stefan Nemanja monument and the surrounding buildings is again in question. The final location solution for Meštrović's Victor in the end proved to be the best choice, and since then this public artwork has been Belgrade's landmark. For Stefan Nemanja, it is still early to say because the monument hasn't yet arrived in Sava square. One thing is sure; the negative reaction is evident, from both the general and professional public.

6.2.2 Monument of Gratitude to France by Ivan Meštrović

Another important public artwork by Ivan Meštrović, also installed in Kalemegdan, is over 4 m high, the Monument of Gratitude to France (Figs. 6.4, 6.5, 6.6, and 6.7). Unlike Victor, this sculpture was fully implemented according to the proposed project and in the planned location. The idea for this project came in 1924 on the initiative of the Society of Friends of France and the Association of former students from French schools, while the Belgrade government donated land for erecting the monument in 1928. The gratitude expressed was for the French military and educational assistance during and after World War I. The spot is central in the beautiful Kalemegdan Park, near the main pedestrian zone of Knez Mihailova Street. Access to the monument is at the end of the main avenue of the park, extending over its boundaries, along Knez Mihailova Street. The Monument of Gratitude to France stands out as a significant visual determinant of the park. Nearby is the French Embassy building (Fig. 6.8). France responded by raising monuments to King Petar I and Alexander I Karadjordjević in Paris (also in Orléans and Marseille). The Monument of Gratitude to France was formally unveiled in 1930.

A female figure cast in bronze rests on a 7-m-high stone pedestal. The artist drew unequivocal rhetoric from the famous symbolic representations of the French Republic, the sculpture Marianne the Goddess of Liberty. Marianne is making a strong step forward in Meštrović's representation. He made an expressive figure in motion, which represents France moving forward to aid Serbia.

The stone base is made of two cubes. On the sides of the lower cube are sculpted allegories in bas-relief, Warriors and the Sorbonne. The ground area is geometrically arranged according to the principles of a French garden. On the front is a carved dedication *A la France*, and at the bottom is the inscription: *We love France as she loved us 1914–1918.* This work brought novelty into public art practice in Belgrade. The Monument of Gratitude to France introduced a new concept of expressive sculpture instead of the previously used realistic concept. It was designated a cultural monument in 1965 and in 1983 a cultural monument of great significance (Cultural Heritage Preservation Institute of Belgrade 2012).

Fig. 6.4 Monument of Gratitude to France by Ivan Meštrović at Kalemegdan. Photographer Biljana Arandelovic. Rights reserved

Some controversies followed the monument at the end of the twentieth century. France took part in the 1999 NATO bombing of Serbia, which as a consequence led to a group of citizens covering the monument with a black cloth and placing a sign with the text: *May there be eternal glory to France which doesn't exist anymore* (Janković 2010).

Fig. 6.5 Monument of Gratitude to France by Ivan Meštrović at Kalemegdan. Photographer Biljana Arandelovic. Rights reserved

Fig. 6.6 Monument of Gratitude to France by Ivan Meštrović at Kalemegdan. Photographer Biljana Arandelovic. Rights reserved

Fig. 6.7 Monument of Gratitude to France by Ivan Meštrović at Kalemegdan. Photographer Biljana Arandelovic. Rights reserved

Fig. 6.8 The French Embassy in Belgrade has a prominent position, located by Kalemegdan Fortress close to the main pedestrian street, Knez Mihailova. This important architectural treasure was built in 1933 by French architect Roger-Henri Expert, and it is one of the most beautiful buildings in art deco style in Belgrade. Photographer Biljana Arandelovic. Rights reserved

6.2.3 *Monument to the Unknown Hero by Ivan Meštrović*

The Monument to the Unknown Hero by Ivan Meštrović is a monumental mausoleum of public purposes on the top of the small Avala Mountain, and it is Meštrović's last implemented monument in Belgrade (Figs. 6.9, 6.10, and 6.11). The mountain is situated 17 km southeast of Belgrade's city center, and its peak offers a good panoramic view of Belgrade. This is a specific location with traces of different civilizations throughout history, in addition to beautiful countryside.

The monument on Avala was inaugurated on June 28, 1938, on the important national holiday of Vidovdan, on the site where the medieval town of Žrnov once stood. Žrnov once constituted an important strategic site, and Avala was a significant mining area. With the arrival of the Roman Empire, Avala gained importance as a mining site where lead, silver, and iron were excavated. The medieval Žrnov was destroyed by dynamite by the order of King Alexander I Karadjordjević in 1934 to free up the plateau for the monument.

The mausoleum was constructed at Avala in the place where in 1922 the peasants from surrounding villages had erected a monument to the unknown fallen warrior. The selection of the location was highly symbolic (Ignjatović 2007). The idea of raising a monument to the Unknown Soldier occurred almost simultaneously in most allied countries after World War I, as a genuine need to express gratitude to the warriors who died for their freedom.

Fig. 6.9 The Monument to the Unknown Hero by Ivan Meštrović. Photographer Biljana Arandelovic. Rights reserved

Fig. 6.10 The Monument to the Unknown Hero by Ivan Meštrović. Photographer Biljana Arandelovic. Rights reserved

The King wanted to show and glorify the idea of Yugoslavia. Since then, this place has been the symbolic location for all future governments to honor all fallen warriors, and for official commemorations of important historic dates. Unfortunately, King Alexander I Karadjordjević was alive just for the occasion of laying the foundation stone, but not to see the work completed since he was assassinated the same year in Marseilles. Erection of the monument began in 1934 and ended in 1938.

The King succeeded in deleting national history by demolishing the medieval town of Žrnov and erecting a monument in honor of the new state, where the supranational idea dominated. At least this was his desire, and as history demonstrated, this was the illusion which he believed until the end of his life. Meštrović's Monument to the Unknown Hero is considered to be the last symbolic attempt of King Alexander I to save the idea of Yugoslavia.

The monument is situated on the top of the leveled mountain, with long access paths leading to the two entrances of the grave. It is built of black granite and conceived as a mausoleum in the classical form of an enlarged sarcophagus. The sarcophagus is situated on a pedestal with six steps. The full Yugoslav symbolism is accomplished by four granite caryatids which are positioned on the portals of the mausoleum. There are caryatids on either side of the entrance, which are figures of women in the national costumes of the Yugoslav peoples. The caryatids at the eastern portal personify the women from Dalmatia and Bosnia, and they are

Fig. 6.11 The Monument to the Unknown Hero by Ivan Meštrović. Photographer Biljana Arandelovic. Rights reserved

Fig. 6.12 Museum of Contemporary Art in Belgrade. Photographer Biljana Arandelovic. Rights reserved

opposite women from Macedonia and Banat, while at the western portal there are Slovenian and Croatian women opposite women from Montenegro and Šumadija.

Landscape, architecture, and sculpture are subordinated to the basic idea of monumentality at this site. All elements are part of an urban entity. The main decorative elements are eight caryatids, which are the main carriers of the Yugoslav concept. On the front of the monument is an inscription, *King Alexander I of Yugoslavia to the unknown soldier.*

6.3 Public Art Within the Museum of Contemporary Art Sculpture Park at Ušće Park

The rich sculpture collection at the Museum of Contemporary Art in Belgrade (Fig. 6.12) consists of 752 works by 196 Serbian and Yugoslav artists, from the beginning of the twentieth century to the present. As part of the collection are works by the founder of modern sculpture in Yugoslavia, Ivan Meštrović: Large widow (1907) and Torso Banović Strahinja (1907) from the Kosovo cycle (on permanent loan from the National Museum in Belgrade). Under the influence of Meštrović, but closer to the aesthetics of academicism is the work of Toma Rosandić, which is also in the Museum of Contemporary Art collection (artwork Harfista from 1934) (MoCAB 2019). He was, along with Meštrović, a prominent Yugoslav sculptor and

Fig. 6.13 Koritarke sculpture by Ivan Sabolić in front of the Museum of Contemporary Art in Belgrade. Photographer Biljana Arandelovic. Rights reserved

Fig. 6.14 Composition 2 sculpture by Oto Logo in front of the Museum of Contemporary Art in Belgrade. Photographer Biljana Arandelovic. Rights reserved

Fig. 6.15 *Composition 2* sculpture by Oto Logo in front of the Museum of Contemporary Art in Belgrade. Photographer Biljana Arandelovic. Rights reserved

Fig. 6.16 Circulation in Space sculpture (Cirkulacije u prostoru) by Vojin Bakić in front of the Museum of Contemporary Art in Belgrade. Photographer Biljana Arandelovic. Rights reserved

also the author of the artwork in front of the House of the National Assembly of Serbia, Horses at Play (placed there in 1939).

The period after World War II was marked by the dominance of late modernism, with abstract sculptures of free forms, the most important representatives in this collection being by Olga Jevrić (who had a pioneering role in this direction with her sculpture Complementary Form (Komplementarna forma, 1956/7), made of freely grouped cement masses of iron reinforcement rods), Dušan Džamonja, and Vojin Bakić. Around the museum building is a spacious park, adorned by the works of the most eminent Yugoslav sculptors of the twentieth century (Figs. 6.13, 6.14, and 6.15). Within this urban setting, as part of the Museum of Contemporary Art complex, in front of the museum building, are two stainless steel sculptures by the influential Yugoslav sculptor, Vojin Bakić: Circulation in Space (Cirkulacije u prostoru, 1970) (Fig. 6.16) and Leafing form (Razlistala forma, 1960) (Figs. 6.17 and 6.18).

Bakić is also the author of some of the most famous Yugoslavian memorials (Stierli and Kulić 2018). For example, his distinguished work at Petrova Gora at the top of Mali Petrovac was officially opened in 1981 as monumental structure dedicated to the partisan and civilian victims of World War II. The foundation stone was laid in 1946. Most memorials like this were commissioned by Josip Broz Tito in the 1960s and 1970s in order to remember the critical battlefields of World War II. During the 1980s, these Yugoslavian monuments attracted millions of visitors each year, especially young pioneers (children aged seven and older) for the purpose of patriotic education. Besides this work at Petrova Gora, other remarkable Yugoslav memorials by Bakić are in Kamenska and Kragujevac. At the beginning of his career, he worked under the influence of the Soviet socialist realism style (Arandelovic 2018b) but later, in the 1950s, he turned toward abstraction and modernist forms, the style in which he created his most recognizable work, Taurus. He mastered his recognizable style through minimalist geometric forms. His masterpieces were constructed with highly polished and reflective stainless steel, the material used for the sculptures in the collection at the Museum of Contemporary Art in Belgrade.

Another significant tendency in sculptural work after 1945 was figurative sculpture; and an eminent artist whose work is installed in front of the museum building, within the sculpture park at Ušće Park is Matija Vuković. His sculpture Miner (Rudar) was made originally in 1962 but was not installed at this location until much later (Figs. 6.19 and 6.20).

The monument to Boris Kidrič (Fig. 6.21) was created by Nikola Janković, who was considered to be one of the best authors of traditional monumental figurative sculpture in Yugoslavia. The sculpture is also part of this outdoor collection and is an example of the role of politics in public art in Belgrade. This is not the only case of this kind of political interference, but definitely one of the most famous examples of it. A 3-m-high monument called the Monument to Boris Kidrič was created as a portrait of one of the chief organizers of the Slovene Partisans and also one of leading politicians in Yugoslavia. It was installed in 1963 to stand permanently on the spot where the monument of Miloš Obrenović now stands (the Prince of Serbia

Fig. 6.17 Leafing form sculpture (Razlistala forma) by Vojin Bakić in front of the Museum of Contemporary Art in Belgrade. Photographer Biljana Arandelovic. Rights reserved

Fig. 6.18 Leafing form sculpture (Razlistala forma) by Vojin Bakić in front of the Museum of Contemporary Art in Belgrade. Photographer Biljana Arandelovic. Rights reserved

Fig. 6.19 Miner sculpture (Rudar) by Matija Vuković in front of the Museum of Contemporary Art in Belgrade. Photographer Biljana Arandelovic. Rights reserved

Fig. 6.20 Miner sculpture (Rudar) by Matija Vuković in front of the Museum of Contemporary Art in Belgrade. Photographer Biljana Arandelovic. Rights reserved

Fig. 6.21 The monument to Boris Kidrič by Nikola Janković in front of the Museum of Contemporary Art in Belgrade. Photographer Biljana Arandelovic. Rights reserved

and founder of the Obrenović dynasty), in front of the Serbian Government building in Knez Miloš Street. The new monument symbolized the Serbian uprising and liberation from the Ottomans in Takovo in 1815. The Monument to Boris Kidrič was relocated in 1997 and is now part of the Museum of Contemporary Art sculpture park. It was created at a specific time to fit into the state ideology (Bubnjević 2011).

A sculpture by Olga Jančić (a student of Toma Rosandić), also set within the Museum sculpture park, examines a closed oval form of organic origin in her work

Fig. 6.22 Core shape II (Jezgro oblika II) by Olga Jančić in front of the Museum of Contemporary Art in Belgrade. Photographer Biljana Arandelovic. Rights reserved

Core shape II (Jezgro oblika II, 1970) (Fig. 6.22). She was one of the founders of Space 8, formed in Belgrade in 1957 and was a relevant figure in the art scene in Belgrade and Yugoslavia. In 2017, she donated her estate of 83 sculptures as a legacy to the city of Belgrade, which is now kept in Heritage House. She belonged to a group of artists who, after World War II, gave an avant-garde impulse to the art scene.

6.4 The Kalemegdan Bridge Project: The Odande Dovde (from There to Here) Sculpture and Pedestrian Bridge from Belgrade Fortress to the Banks of the Sava

The Kalemegdan Bridge project is a result of teamwork between two renowned artists, Richard Deacon, a British abstract sculptor and Turner Prize winner, and Mrdjan Bajić, a contemporary Serbian sculptor and professor, whose sculptures marked the first historical presentation of Serbia as an independent country at the Venice Biennale in 2007, after the Yugoslav disintegration. Bajić is also the author of the Zoran Djindjić memorial, another ongoing contemporary public artwork. The Kalemegdan Bridge project combines abstract art with new functionality in the form of a massive sculpture adjacent to the bridge. The height difference between the fortress and the Sava River coast was a big challenge for the artists.

The gangway is a 75-m-long concrete/metal structure (width 2.5 m), with two further limbs which are separated from the main axis. One leads to a car park (which is located on the roof of Beton Hall) and the other leads to the park next to the steps. As the gangway crosses a large difference in altitude, the descent will be enabled by means of a long sinuous ramp which stands next to Beton Hall. The spiral ramp is longer than the entire walkway, with a length of 82 m; it carries a range of 11 different columns. An elevator tower is part of this massive construction, and it forms the podium for the artistic sculpture (Deacon and Bajić 2013).

The Kalemegdan Bridge connects part of Knez Mihailova Street and Belgrade Fortress to the coast of the Sava near the Beton Hall complex by means of a safe crossing. The project will significantly facilitate the movement of visitors. At this point, there is no direct access from the fortress to the coast because of the existing urban frames, and there is heavy traffic on Vojvode Bojovića Boulevard and the railway next to the fortress. This bridge will allow pedestrians to reach the banks of the river in 12 min.

Richard Deacon started partnership on this project with Mrdjan Bajić during his first stay in Serbia in 2006, when he was part of a cooperation project between the National School of Fine Arts, Paris and the Faculty of Fine Arts, Belgrade. During Deacon's Belgrade visit, they were invited by the management of the Belgrade Fortress to jointly develop a proposal for a sculptural space in the fortress. The pedestrian bridge project from Belgrade Fortress to the banks of the Sava was initiated the same year, 2006, and it has been a long journey, with its completion due in 2019 (Figs. 6.23 and 6.24).

Cooperation between the Belgrade Fortress Public Company, Henry Moore Foundation, Handel street projects, and the Beacon Art Project from Great Britain has enabled the completion of the first phase of the Kalemegdan Bridge. The artists started in 2009 on a topographic model at the Faculty of Fine Arts Belgrade. Drafting of the conceptual design for the footbridge continued, thanks to workspace offered for the project by the Museum of the City of Belgrade in the premises of the Military Printing building in Resavska Street. In the following years, the necessary urban planning documentation was provided and the city government gave its support. In 2009, the project was presented to the public in an exhibition. When the exhibition finished, the prefabricated construction model was moved to Belgrade Fortress, near the locations planned for the project, where it stood during the summer and autumn of 2009.

Work on the technical documentation continued in 2010 in cooperation with the Faculty of Architecture in Belgrade. After a 2-year break, in 2012, the project authors and the Belgrade Fortress Public Company, its initiator, continued to work on its implementation. According to Deacon, co-author of the project, the bridge enters the fortress, then crosses the border, and at the point where it is closest to the river, it represents an entry portal.

An essential part of the project is the monumental sculpture Odande dovde (From there to here) that is located on the bank of the Sava, next to the Beton Hall complex (Fig. 6.25). Mrdjan Bajić explained that the sculpture evokes an encounter

Fig. 6.23 Construction of the pedestrian bridge from Belgrade Fortress to the banks of the Sava. Photographer Biljana Arandelovic. Rights reserved

Fig. 6.24 Odande dovde sculpture (From there to here) during construction along Beton Hall complex. Biljana Arandelovic. Rights reserved

Fig. 6.25 Odande dovde sculpture (From there to here) in construction with the view of Kalemegdan in the background. Photographer Biljana Arandelovic. Rights reserved

between chaos and geometry, natural and urbanized, modern and traditional, water and city, within unstable historical circumstances, and the particularities of the geographical environment. The sculpture has two parts. The upper part of the sculpture, made of stainless steel by British artist Richard Deacon, was brought from England, while the lower part, made by Mrdjan Bajić, is cast aluminum that was produced in Serbia. The whole sculpture weighs about five tons and is almost 7 m high (Figs. 6.23 and 6.24).

The artists have donated the Kalemegdan Bridge project to the City of Belgrade, while the making of the massive sculpture Odande dovde (From there to here) was completely funded by the Delta Foundation (a private Serbian foundation), which then donated it to the city of Belgrade. Endowment was an important part of Serbian tradition, especially in the period from the end of the nineteenth century until World War II. Today, it is quite rare, unfortunately. This is the largest donation of private capital in the field of visual arts in modern times. If Belgrade wants to keep up with modern times, this practice should continue in the future, with private capital becoming more engaged in developing the cultural scene in Serbia. This implies that the state, above all, should be much more involved in developing public art practices. "*Public art can also have the role of increasing the level of public awareness and bringing contemporary art closer to the masses; then, it develops the symbiosis between town planning and art, raises the potential for tourism, and increases cooperation between town planners, architects and artists, and the administrations of cities*" (Arandelovic 2018a: 29). After a long time, this is the first work from the field of contemporary visual arts with full visibility in Belgrade.

6.5 Sound Installation and Arrow Sculpture for the Zoran Djindjić Memorial

There are different kinds of memorials in the urban spaces of contemporary cities. Memorials confront us with the decision about how we will treat them. They are present to remind people of decisions taken in the past and the consequences of them (Arandelovic 2018c). Understanding the choices we make and their possible influence on our future decisions is crucial for understanding the conceptual sound installation and the idea behind the Zoran Djindjić memorial. The paraphrased famous quote of assassinated Prime Minister, "look into the future, because there is where you and I will meet" is the guiding motif behind the project. The monument is in the shape of an arrow pointing upward, moving toward the future.

Memorials are a sensitive issue in a country like Serbia. The whole situation surrounding the memorial sculpture of assassinated Prime Minister Zoran Djindjić is complex. There has never been more discussion about a sculpture than about this one, although the Stefan Nemanja monument is also very topical. Zoran Djindjić was the Prime Minister of Serbia from 2001 until his assassination in 2003 and he was a long-time opposition politician. He was the first prime minister after the fall of Slobodan Milošević and for some he was a great reformer, while for others a traitor. This is one of the reasons for this great controversy but another reason is the fact that the government that organized the competition represents the politics Zoran Djindjić was clearly against and fought against during his life.

The author of the winning project for the memorial competition was Mrdjan Bajić, with associate sculptor Predrag Keśelj, architect Petar Sazdanović, art historian Marina Andrić, and playwright Biljana Srbljanović, who designed the sound installation, along with sound designer Dobrivoje Miljanović and programmer Vladimir Riznić. The results were released in 2017 and the construction project is ongoing, the first of its kind in Belgrade—monumental, modern, and abstract (Figs. 6.26 and 6.27).

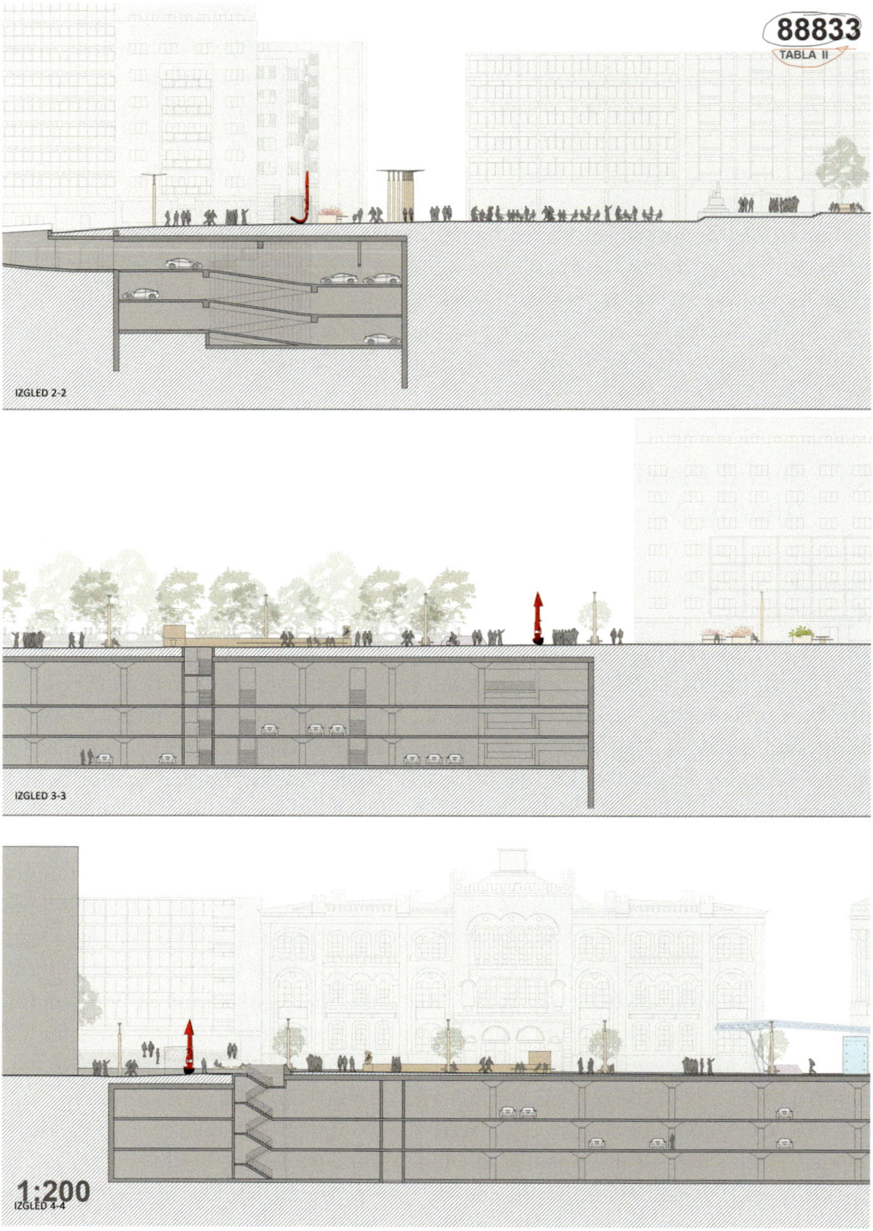

Fig. 6.26 Arrow, the winning project for the Zoran Djindjić memorial. Image provided by the City of Belgrade Urban Planning Office

The position for the future monument, near the building where Djindjić lived and faculty where he studied (Faculty of Philosophy), was determined by Austrian architect Boris Podrecca, who was the head of the five-member competition jury

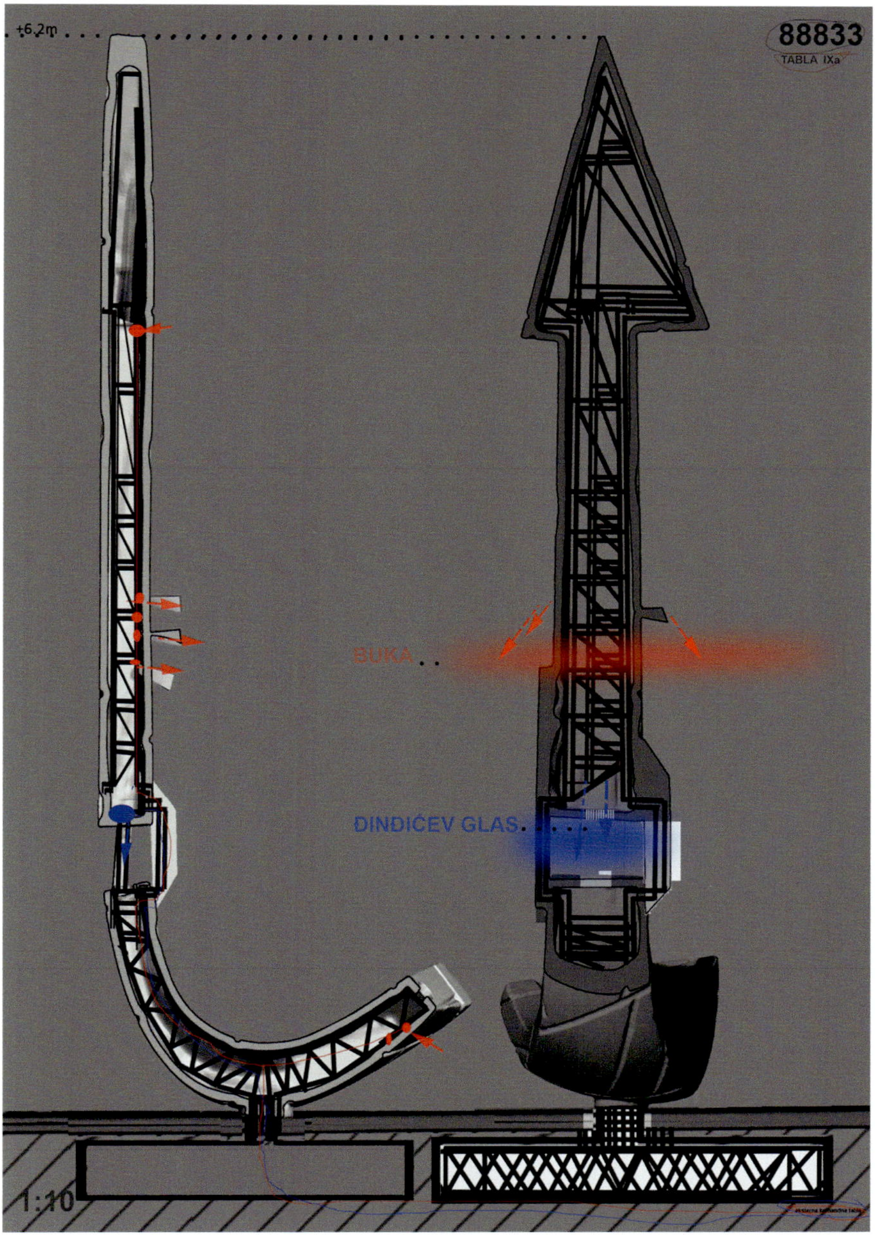

Fig. 6.27 Arrow, the winning project for the Zoran Djindjić memorial. Image provided by the City of Belgrade Urban Planning Office

and also the architect who designed the reconstruction of Student Square (Studentski trg). This important central square is currently in the process of reconstruction (more in the chapter Sustainable Belgrade. Expansion of the

Fig. 6.28 The Rectorate of the University of Belgrade at Student Square. Photographer Biljana Arandelovic. Rights reserved

pedestrian zone in the city center). Student Square is adjacent to the Academy Park and connected with the main central city pedestrian zone of Knez Mihailova Street. It is in the vicinity of Republic Square and also has some important historical buildings. The Rectorate of the University of Belgrade is situated in Captain Miša's Mansion, built in 1863; it was declared a Monument of Culture of Exceptional Importance in 1979 (Fig. 6.28). Student Square number 5 houses the Kolarac People's University Building with a concert hall, which is also a monument of great cultural and historical importance, declared as immovable cultural property due its irreplaceable cultural role and remarkable architectural value. The Ethnographic Museum is situated in this square. In this way, the Zoran Djindjić memorial will occupy one of the most prominent locations in the city center.

The modern history of Serbia is the context in which this project occurred. The time in which Zoran Djindjić lived but which he did not outlive is full of many traumatic events, all with different interpretations. Thus, the noise created around this memorial is a logical consequence and at the same time it also creates the basis for understanding the trauma. The trauma is circulating around. It was present in the political events that preceded his rise to political power, and culminated in the moment of his assassination in 2003; it has not disappeared even 16 years after his tragic death.

Sound is an important element of this project. The memorial's authors use loud volume to emphasize the importance of Djindjić's words. The streets have always been flooded with some kind of noise like shouting, cheering, and roaring. It is not always possible to hear the right words because of unnecessary noise. This monument dedicated to Zoran Djindjić has thus become a reminder to future generations of the existence of this philosopher and politician, because it is a sound installation that broadcasts his famous quotes and sentences. This memorial is becoming a space of remembrance, as well as a reminder of collective Serbian trauma and of the fact that nobody actually heard what he was saying.

The sculptural work, in the form of a gigantic arrow, is a place to hear, not just to watch. The body of the sculpture is 6 m in height. It aims to provoke reasonable thinking. The sound monument starts a conversation with passers-by, visually, and mentally. With this, the passer-by becomes an integral part of the project, sculpture, and idea. It aims to provoke individuals to reflect on their role in society. The form of the monument is a timeless visual symbol, an arrow that directs the direction of movement. This arrow is physically broken; there is one part missing.

The direction is still there; the arrow continues toward the sky (the future) but there is no physical contact with the base of the arrow. At the same time, it is both broken and firm. The indication is clear, and the arrow is broken but it continues its way up in the sky; the direction is still the same even the structure is physically interrupted. The arrow is a universal symbol for direction.

The essential questions raised by this memorial and sound installation are as follows:

What have I done?
What can I do?
What haven't I heard, understood?

Could I have made a difference?
What have I failed to listen to and understand, and why has this happened to me?
Why can't I do anything now?
How can I do something now?

The authors of this contemporary public artwork firmly believe that space for the memory of Zoran Djindjić must be from the field of contemporary art built on the postulates of contemporary visual art, as his ideas were in the modern spirit, always looking toward the future.

The selection of words and thoughts by Zoran Djindjić includes some of the words he spoke when he was not in the midst of daily political challenges. This testifies that he was a great thinker and analyst of his time, an infallible observer of human nature and of the people with whom he lived. Part of the sound installation is, in the words of Djindjić: *Tell people the truth; you need to wash with cold water; if this does not help, take a shower with cold water; clear your house, literally and metaphorically; Life is a miracle given to us; you will sleep when you retire* (Bajić et al. 2017).

This installation has not reached the implementation phase since 2017; when the competition results were announced, it is no longer a topic. The conceptual design for the monument to the assassinated Prime Minister Zoran Djindjić was presented in 2017, and its placement in Student Square was scheduled for 2018. The official explanation is that the erection of the memorial was rescheduled because of the current construction works at Student Square, and it has had to be postponed for 2021, when the planned full completion of all construction work is scheduled. It remains to be seen what will happen to the memorial and whether it will actually be built, or whether it will end up being one of the many planned projects in Belgrade that has never materialized.

6.6 Stefan Nemanja's Statue and Historical Theme

Mrdjan Bajić, professor and sculptor and one of the most successful contemporary artists in Serbia (Kovač 2017), gave his opinion on the current situation in terms of art in public space in Belgrade, pointing out his fear of the sculptures which today are about to be raised in Belgrade's public space. According to him, these sculptures look more like those erected in Germany and Italy in the 1930s and 1940s, and in the 1950s in the USSR, than what is today constructed in a modern European metropolis. As an example of such a tendency, Bajić highlights the winning solution for the Stefan Nemanja monument competition (Čaluković 2018).

Parallel with the development of the Belgrade Waterfront project, new public artworks are being created in Belgrade. The most monumental of these is the monument of Stefan Nemanja, a traditional figurative representation of the most important historical figure for the Serbian people. This will be the highest monument in Belgrade. After its construction, it will be the largest structure of this type in the whole country.

Nothing is apolitical in Serbia, especially the construction of a monument of such dimensions, so it is not surprising that the proposal for this massive (28 m) monument of one of the most important medieval Serbian rulers (Fig. 6.29) and founder of the country came directly from the Serbian president, Aleksandar Vučić. The tender documentation stated that it was necessary to provide a solution for the monument and its close surroundings. It was additionally stated that the figure representing this historical figure from the twelfth century must be in the spirit of realism and a figure of this medieval ruler presented in his middle age.

Fig. 6.29 Stefan Nemanja's monument model. Image provided by the City of Belgrade Urban Planning Office

Stefan Nemanja's monument will be located in the middle of the new Sava Square in the Savamala district, in front of the historical building of the railway station, which will become the Museum of Serbian History. Belgrade Waterfront is in the background, in the immediate vicinity, part of the new urban identity (Arandelovic et al. 2017) that is currently under construction.

The winner of the international competition for the construction of Stefan Nemanja's monument is Russian sculptor Alexander Rukavishnikov, who teamed with Serbian architect Petar Arsić, who was in charge of the monument's surroundings. The required specification was for it to be a figurative composition in standing position. The ruler, Nemanja, is holding the Orthodox cross up high. The statue has no ordinary classical pedestal. The pedestal is designed as a cracked Byzantine helmet, from which the ruler's scepter emerges. Passers-by will have the possibility of entering the helmet, presented in the form of a canopy with a series of reliefs with depictions of medieval Serbia inside it.

The ambitious monument of Stefan Nemanja is scheduled to be revealed in the first half of 2019, marking 800 years of the autocephaly of the Serbian Church. The monument will be built in Russia, and then transported to Serbia. It will have a representative, central place in the city. Observing this example, it is evident that Belgrade, as the capital of Serbia, is returning to solutions from the past, instead of keeping pace with the world and the idea of a modern metropolis. The current government has announced a continuation in the construction of monuments of prominent historical figures.

The opinion of Serbian historian Dejan Ristić is that the initiative to raise a monument was good but he is criticizing this solution. He believes that it is very important that the founder of the Serbian medieval state gets a monument in Belgrade. In his opinion, the credibility of the jury is was questionable, since in it were only two sculptors and not even one historian specializing in the Middle Ages (Jovičić 2018).

During the selection process, jury member, sculptor, and professor Milorad Živković resigned and accused a politician (also a jury member) that he had imposed his opinion by favoring the Russian sculptor. His explanation is interesting, since he said that the jury had asked the sculptor to reduce the dimensions of the sculpture due to the size of the buildings in the immediate environment and their mutual relations. It should be emphasized that the railway station is a building of a great cultural importance that was declared a cultural monument in 1981, and heritage of great importance in 1983. A sculpture of this size may swallow it. This objection was not adopted by the author (Živković 2018). On this basis, there is a suspicion that politics influenced the final solution. This monumental visual solution fits better into Russian but not Serbian society and the urban framework of Belgrade. Public space is a place where different heterogeneous social groups become visible to each other and openly display their identity. It is where excluded groups gather to seek legitimacy from the dominant public culture. Public space is a deeply political ground; it is the location of inclusion and exclusion, of visibility and invisibility (Poposki and Todorova 2016). Belgrade is trying to keep up with the developed world, thus implementing more public artwork in its open spaces.

Unfinished reform and institution-building, on one hand, and ambition, on the other hand, emphasize the need to find an authentic model that will help to clarify problems (Arandelovic et al. 2017). There are many unsolved problems regarding public art practice. The inclusion of a wider circle of experts is a step forward. Maximum and clever use of all the resources in the field of art and culture is essential. Public hearings, the presentation of ideas, exhibitions, and useful discussions are necessary in Belgrade. Belgrade should learn from cities like Berlin and Graz, where such practices are at the highest level.

If we take Graz as a positive example, with its Light Sword statue, a work by Austrian artist Hartmut Skerbisch, erected in front of the Opera House in 1992 for the occasion of the Steirischer Herbst festival, we can notice how a monumental statue of 54 m in height stands in the historical central city core without destroying the urban setting. On the contrary, this work looks into the future while enjoying its traditional architectural setting. The Austrian artist constructed the Light Sword statue in Graz, inspired by the celebration of 500 years since the discovery of America; he based his work on the Statue of Liberty in New York, making it the same height (Arandelovic 2015).

6.7 Concluding Words: What Belgrade Has to Offer

6.7.1 Belgrade's Sculptural Masterpieces from the Period Between the Two World Wars

The most fruitful period for art in public space in Belgrade was between the two wars, when the most important public monuments found their permanent positions, and many of them are today the symbols of the city. Among these we include Ivan Mestrović's sculptures—Pobednik (Victor), the Monument of Gratitude to France and the Monument to the Unknown Hero—as well as the monumental Horses at Play, a bronze composition in front of the House of the National Assembly of Serbia (1938) by Toma Rosandić, which is one of most iconic artworks in public space in Belgrade, on King Alexander Boulevard. Toma Rosandić and Ivan Meštrović were two of the most prominent Yugoslav sculptors, and rivals, both working in the same period and shaping the art scene between the World Wars. A member of the Serbian Academy of Science and Arts, Rosandić, established a prominent school in Belgrade known as the Master Workshop, before producing his greatest masterpiece, Horses at Play (Igrali se Konji Vrani), a pair of statues of muscular male characters struggling with two raging horses (Figs. 6.30, 6.31, and 6.32). He was born in Split (Croatia) but he spent most of his productive years in Belgrade and was known, as was Meštrović, as a true Yugoslav.

Rosandić never gave an official interpretation of the symbolism of this composition, and the official explanation was that the work is an artistic presentation of sculptural forms and how human and animal bodies interact. There are several

Fig. 6.30 Horses at Play sculpture by Toma Rosandić in front of the House of the National Assembly of Serbia. Photographer Biljana Arandelovic. Rights reserved

Fig. 6.31 Horses at Play sculpture by Toma Rosandić in front of the House of the National Assembly of Serbia. Photographer Biljana Arandelovic. Rights reserved

Fig. 6.32 Horses at Play sculpture by Toma Rosandić in front of the House of the National Assembly of Serbia. Photographer Biljana Arandelovic. Rights reserved

interpretations of symbolism in this composition, but the general view is that the figures represent the common men who stand powerless in front of the government, struggling with a stronger force, authority. Some believe that the figure with a horse on his hunched back is an allegory of oppressed people, while the second figure is an allegory of a man who has fought with the force, as a symbol of resistance (Đorđević 2016).

Many memorials in Belgrade were erected in the period between the two world wars, but after the end of World War II, especially during the 1980s and 90s, a large number of new memorials were dedicated to the prominent military leaders from the First World War (Božović 2014). There are a large number of monuments dedicated to historical figures throughout the urban core of Belgrade (Fig. 6.33), in squares, parks, and streets, and this is the dominant way of displaying art in public space in Belgrade.

6.7.2 Street Art and Murals

The situation regarding art in public spaces is in line with the general situation in Belgrade. The crisis in Serbian society has been followed by a crisis in its art, which was especially emphasized during the 1990s. The atmosphere of hyperinflation and sanctions had a huge impact on the art scene, influencing the street art scene in

Fig. 6.33 The Karadjordje monument is located in front of the St. Sava Temple and the National Library in Vračar in Belgrade. This bronze monument by Sreten Stojanović was built in 1979. It has a total height of 3.2 m. Karadjordje was a Serbian revolutionary who led the battle for independence from the Ottoman Empire during the First Serbian Uprising. Photographer Biljana Arandelovic. Rights reserved

Fig. 6.34 Dorćol quarter building with graffiti. Photographer Biljana Arandelovic. Rights reserved

particular. Under these conditions, the street art scene in Belgrade became more active and developed its special form, and street artists like Artez, Junk, Rage, and TKV helped to position Belgrade on the international scene. Street art has emerged over the whole city, but it is more dominant in the Savamala, New Belgrade, and Dorćol districts. Since 2017, tourists have been able to have an alternative guide experience of Belgrade with Street Up Tours, which are organized by local street artists, with a focus on the Dorćol (Fig. 6.34) and Savamala areas (Djordjević 2016).

Murals have a tradition in Belgrade, and not only one which is linked to the alternative scene that has gained importance with the development of Savamala's creative district. The tradition of painting Belgrade's facades with murals originates from the 1970s, thanks to the initiative of Professor Čedomir Vasić from the Faculty of Fine Arts Belgrade, in cooperation with the Cultural Center of Belgrade.

The first known mural, unfortunately destroyed, was painted in Belgrade in 1970 by Lazar Vujaklija on the facade of a small craft shop on King Alexander Boulevard. Within the framework of Latin America Week in 1977, an art group called Brigade Salvador Allende from Chile painted a mural on the walls of the Student Cultural Center (SKC) with students from the Faculty of Fine Arts, called Solidarity of the Yugoslav peoples with the peoples of Latin America.

Another important initiative in connection with murals was related to the IX Summit of the Non-Aligned Movement, held in Belgrade in 1989. For this occasion, five murals were completed, one of which was the work of Vladimir

Veličković, an internationally acclaimed Serbian painter, was on the facade of the Faculty of Philosophy. The Institute for the Protection of Monuments was given the responsibility of organizing the murals and the artwork was led by Slobodan Jevtić-Pulik.

In 1982, the Center for Urban Planning and Development of Belgrade organized a conference entitled Communications, where for the first time the question of murals as a way of decorating facades in Belgrade was raised. In the following year, 1983, Professor Čedomir Vasić, together with his students, submitted a map with 50 conceptual mural designs to the then Mayor, Bogdan Bogdanović, after which a Council of the Belgrade Assembly adopted the Operational Plan for the Development of the City in 1989. Unfortunately, the years of crisis and destruction followed, and none of this materialized.

Among recent initiatives are Belgrade Summer Festival (BELEF) and the organization Kiosk—Platform for contemporary art. The results of a research project, the *History of mural painting in Belgrade*, which began in 2011 and was implemented by Jovan Stojanović, Miloš Miletić, and Mirjana Radovanović, showed the rich history of facade painting in Belgrade, but also indicated that many murals had been destroyed or damaged. Various factors caused this, but the ignorance and indifference of the community played a decisive role. In most of these projects, local communities had not been involved in decision-making or in the production of the public art projects, thus for the population, they were not visible as public art (Stanojević et al. 2019).

6.7.3 Perspective

There are very few examples of contemporary art in public space in Belgrade today. The monumental sculptures From there to here, by Mrdjan Bajić and Richard Deacon, is the only one of its kind in Belgrade, although the entire Kalemegdan Bridge project, which this sculpture is part of, has not yet been fully completed. The sculpture was finished and placed on the site by the river Sava, but the rest of the construction project, the footbridge, created controversy when members of the public raised their voice over the level of construction costs. Although it should already be completed, the project still has to wait, and the works are currently suspended. When completed, this will be the first project in the field of contemporary art in the center of Belgrade.

Currently, there are no plans for a new public artwork project that could have a greater impact than Meštrović's Victor in Kalemegdan and become the new symbol of Belgrade. Belgrade does not yet have a modern variant of Victor, and so a monument erected in 1911 is the main landmark of Belgrade. Belgrade needs to put more effort into this field if it is to reach European and world metropolis standards. A city with such a cultural tradition can certainly do much more to make up for time lost during the final decade of the twentieth century.

Construction of the monument to Stefan Nemanja is an attempt to create a new monumentality. However, a monument of this kind is outside of modern trends and has come at the wrong time. Also, the initiatives and decisions about the author of the monument were made politically and not by professionals. This attitude cannot bring positive results, but only irreparable damage. However, the gigantic statue of Stefan Nemanja has not yet taken a permanent position on the new Sava Square so it is too early to pass a final judgment. The project is currently in the implementation phase. The professional public has protested but it has not changed anything.

In Belgrade, there are many more urgent and major issues of art. Once politicians realize that art is important and necessary for the construction of modern society, for strengthening its identity, educating the population, and improving culture, then Belgrade will be a more competitive world capital.

Belgrade has potential, cultural institutions, and museums with a tradition and reputation; it also has art schools (Academy of Fine Arts and the Faculty of Applied Arts) and enough empty locations for the development of art in public space. All it takes is the willingness of politicians to listen to those who know much more than them in this field. Luckily, the Milošević era didn't leave a visible mark on the monumental culture of Belgrade, because such errors would be difficult to repair. Although it left significant negative consequences on the urban city core, primarily by way of illegal construction and illegal buildings, it thankfully made no large visible changes relating to art in public space.

Belgrade is now developing in the right direction, and the city is returning to the international scene. Even the world's media, including the BBC and the Guardian, are recording features, thanks primarily to the development of Savamala as a creative district. Belgrade does not keep up with global metropolises in terms of projects related to art in public spaces. It is essential for those who make decisions to understand what features and benefits of this field they can bring to the city. Ultimately, public art is more than just art; it is also a stimulant for creating new businesses and furthering the development of the creative economy.

References

Arandelovic B (2015) Graz, UNESCO city of design and historical heritage. Cities 43:78–91

Arandelovic B (2018a) Introduction to public art. In: Public art and urban memorials in Berlin. Springer International Publishing, pp 28–38

Arandelovic B (2018b) Socialist realism in art. In: Public art and urban memorials in Berlin. Springer International Publishing, pp 257–277

Arandelovic B (2018c) Urban memorials. In: Public art and urban memorials in Berlin. Springer International Publishing, pp 279–362

Arandelovic B et al (2017) Belgrade: imaging the future and creating a European metropolis. Cities 63:1–19

Bajić et al (2017) Competition project for the Zoran Djindjić memorial

Borić T (2005) O istoriji i arhitekturi Terazijske fontane [About the history and architecture of the Terazije fountain] Nasledje (6):227–236

Božović A (2014) Memorijali Prvog svetskog rata na teritoriji Beograda, Beograd. Zavod za zaštitu spomenika kulture grada Beograda Cultural Monument Protection Institute
Bubnjević S (2011) Tito u podrumu akademije. Vreme
Čaluković N (2018) Istorija će biti surova prema našoj generaciji [History will be harsh towards our generation]. Nedeljnik. 26 April 37–41
Deacon R, Bajić M (2013) Susret. Most na Kalemegdanu [Encounter. Kalemegdan Bridge]. JP Beogradska tvrdjava
Djordjević A (2016) Street Art Belgrade, Komshe
Đorđević M (2016) Šta ne znamo o konjima vranim ispred Skupštine. Politika online 4.2.2016
Ignjatović A (2007) Jugoslovenstvo u arhitekturi 1904 – 1941. Građevinska knjiga
Ignjatović A (2014) Images of the Nation Foreseen: Ivan Meštrović's Vidovdan Temple and Primordial Yugoslavism. Slavic Rev 73(4):828–858
Janković Z (2010) 80 godina Spomenika zahvalnosti Francuskoj - A la France! [80 years of the monument of gratitude to France—A la France!] Vreme
Jovičić U (2018) 28 metara Stefana Nemanje [28 meters of Stefan Nemanja]. Nedeljnik. 14. February 11
Kovač L (2017) Mrdjan Bajić – Rasprizorujuće transverzale [Mrdjan Bajić: disenacting transversals]. Supervizuelna
MoCAB Museum of contemporary art Belgrade (2019). http://www.msub.org.rs/zbirka-skulpture
Pavlović M (2016) Skulpture kao toponimi Beograda - Ivan Meštrović, Project Nepokretno nasledje u Srbiji [Immovable heritage in Serbia]
Poposki Z, Todorova M (2016) Public memory in post-conflict Skopje: civic art as resistance to narratives of ethnicity and disintegration. Post-Conflict Performance, Film and Visual Arts 95–112
Stanojević J, Miletić M, Radovanović M (2019) Istorijat oslikavanja murala u Beogradu. http://www.udruzenjekurs.org/en/
Stierli M, Kulić V (2018) Toward a concrete Utopia: architecture in Yugoslavia, 1948–1980. The Museum of Modern Art New York
Stojanovic M, Stanojev I (2016) The quest for the national identity of the Kingdom of Yugoslavia: from the Vidovdan Temple to St Sava's Church. Nasledje
Vukotić Lazar M, Danilović Hristić N (2012) The significance of contemporary art in process of the interdisciplinary design of the public urban space of Belgrade. Zbornik 8/2012. Museum of Applied Arts Belgrade, 119–129
Živković M (2018) Liči na cara Dušana, ali nema veze, oba su Nemanjići [It looks like the Emperor Dušan, but it does not matter, they are both Nemanjići]. Danas 12 May

Chapter 7
Pedestrian-Friendly Belgrade

Milena Vukmirovic

Abstract During the 70s, Belgrade was developed in line with other prominent European capitals. This particularly refers to the creation of pedestrian zones in the city center similar to those that at that time appeared in Paris, Munich, and Vienna. The idea of creating a pedestrian zone in the center of Belgrade emerged in the early 70s and was formally defined in the General urban plan from 1972, and later in the Regulatory Plan for the Central Zone of Belgrade from 1977. The first pedestrian zone along Knez Mihailova Street was opened to pedestrians in October 1988. By mid-2010, the majority of the planned interventions which had contributed to forming the pedestrian zone in the center of city were finished, and the next phase of extending it to the river waterfronts began. Following the recommendations for sustainable urban development and the introduction of a new green hierarchy of urban transport, current development trends in Belgrade favor the further expansion of the pedestrian zones to the banks of the Sava and Danube. This chapter presents an overview of past and current efforts to apply the principles of sustainable development and create pedestrian-friendly areas in Belgrade in the area defined as Belgrade Central Zone. This zone covers four city cores: the central core of Belgrade, the central core of New Belgrade, the central core of Zemun and Sava amphitheaters. In accordance with that, the chapter features important ongoing projects and shows the development process of pedestrian areas in Belgrade, as well as the sustainability of the concept from the 70s.

The quality of life and the quality of the environment in cities are notably affected by road transport and traffic. Considering the main problems associated with increased urban traffic and congestion, negative impacts on the urban quality of life can be found in the following domains: visual instruction, noise and vibration, energy consumption, severance, competitiveness, accidents, air pollution, loss of urban "living space," economic efficiency, and equity (European Commission, Environment Directorate-General 2004). Following the statistics and some predictions, 80% of Europeans live in an urban environment; every year more than three million cars are added to the car fleet in Europe, and the total road traffic kilometers in urban areas will grow by 40% between 1995 and 2030. As the issue

© Springer Nature Switzerland AG 2020

B. Arandelovic and M. Vukmirovic, *Belgrade*, The Urban Book Series,
https://doi.org/10.1007/978-3-030-35070-3_7

has been identified, there is a need to find a solution that could on one hand increase the road space available for cars, while on the other hand, reduce the car capacity as a sustainable, efficient planning solution. The second solution could improve the quality of life and quality of the environment in cities, with special attention given to the introduction of pedestrian zones and the creation of a pedestrian-friendly environment.

The framework made in order to solve these problems is The European Sustainable Urban Mobility Plan (SUMP). It was developed on the basis of the broad exchange between stakeholders and planning experts across the European Union in order to describe the main features of sustainable urban mobility and transport. As a methodology recommended by the European Union, most cities have changed their concepts of urban development by reconstructing their centers and entire urban transport systems. In addition, the policy of changing the modes of transport in cities means more space for pedestrians, bicycles, and public transport, and a reduction in the level of comfort for passenger cars. This situation is perfectly explained in the Manual for Streets (Department for Transport 2007) as a hierarchy of traffic users, with the purpose of assisting in decisions related to design, planning, and development. The hierarchy places pedestrians at the top, followed by cyclists, and then public transport, with unaccompanied private car users last.

Following the principles of SUMP and the transport hierarchy, the goal is to change the function of transport for longer distances from cars to public transport (metro, tram, bus, etc.), while for short distances to encourage the use of walking and bicycles. In addition to this, SUMP proposes using modes of transport based on energy from sustainable sources (electrical buses, electrical cars, etc.) and the introduction of a car-sharing system for those who need to make short car journeys in the city, as well as park and ride, which combines using a car for traveling out of the city and public transport for movement in the city. One of the most recognizable manifestations of SUMP is the plan to implement car-free strategies, which can be recognized in many European cities.

Although the idea dates back to the 70s, city plans to go car-free have been very high profile in recent years. In 2014, several European cities announced their ambitions to convert parts of their central districts into pedestrian-oriented places with less automobile congestion and air pollution. They include Paris, Madrid, Oslo, Milan, Dublin, and Brussels, while Belgrade also announced an initiative to extend the pedestrian zone in the center of the city in mid-2015.

Concerned about the worsening air quality in the center of the city, the City of Paris announced the project "Paris Respire" (Paris Breathes) in order to decrease the level of air pollution in its central area (Airparif 2014). The project has defined a car-free scheme for several central locations (the River Seine, le Marais, The Canal Saint-Martin, and Montmartre), where certain roads are closed to vehicular traffic on Sundays and public holidays. Cycling and walking are the main forms of getting around at those times. Along with this initiative, which has recorded significant improvements with regard to the air quality, the city authorities are developing a new pedestrian area along the Seine and expanding the 30 km/h zone. In 2012, the City of Madrid adopted a revised version of its General Urban Plan, with the aim "to transform much of a car-snarled city center into a leafy pedestrian's dream" (O'Sullivan 2013). The Plan foresaw the transformation of 24 major streets in Madrid by removing car lanes, adding bicycle lanes and trees, and putting a new hierarchy in place in which pedestrians come first, then public transport, then bicycles, and then cars. Additional steps to keep inner Madrid car-free were also suggested, with strictly limited parking places in central buildings to encourage people onto public transport. Dublin also introduced an urban plan that aimed to ban cars from parts of the city center by 2017. With these kinds of measures, the authorities are expecting "to reach a point where one in five people commutes into the core via private car, while the rest come by transit (55%), bike (15%) or on foot (10%)" (Eric 2015). Similar to Dublin, the City of Oslo has also announced a plan that has the goal of going car-free, banning private cars in the central district by 2019. The city will enhance its bike infrastructure by building "at least 60 km of bicycle lanes and give a 'massive boost' to public transit" (Fouche and Solsvik 2015). The "total pedestrianization of the historical centre" is the aim of the City of Milan's local authorities. Even if they do not have a clear timeframe for this action, this could be seen as a continuation of activities that started with the introduction of an effective congestion pricing scheme in the core of the city. Known as Europe's most congested city, the City of Brussels has a plan to turn the central axis into a pedestrian zone (O'Sullivan 2014). That will almost double the existing pedestrian zone (from 28 to 50 ha, see Fig. 7.2). The implementation of this vision will cause Brussels to have a slightly smaller pedestrian zone than Venice.

Similar to the European cities mentioned, the City of Belgrade's local authorities are working on extending the pedestrian zone of Knez Mihailova Street, based on initiatives announced in mid-2015 and adopted within the Project IME: Identity_Mobility_Environment of the City of Belgrade, including "the implementation of multiple projects in order to further develop the pedestrian zone in the heart of Belgrade, which started in 1980" (Folić and Vukmirović 2015). In addition to this project, the chapter aims to present initiatives, projects, and realizations that include a somewhat wider central area, known as the Belgrade Central Zone, in order to point to the sustainability of an idea created during the 1970s that can still be interpreted in a contemporary way.

7.1 Belgrade Central Zone and Its Cores

In the late 1960s and early 70s, parallel with the development of the idea of an independent metro system, came the idea of building pedestrian zones. This period was characterized by a dynamic increase in individual motorization in the city, which was one of the key issues that needed to be addressed. To illustrate this, the number of individual registered vehicles increased from 40,438 in 1965 to 300,420 in 1975.

It was considered that the realization of an autonomous, non-surface transportation and communication system had the goal of providing the necessary and additional conditions for organizing nonautonomous surface systems in central city zones. More precisely, this was considered as a precondition for the creation of pedestrian zones. The idea of locating pedestrian zones on both sides of the Sava River within Belgrade's central zone was also in line with the City of Belgrade's urban development, defined in the General Urban Plan for 2000.

Belgrade's central zone is made up of a green center on the confluence of the rivers, with three central nuclei (the old core of the city, the core of New Belgrade, and the core of Zemun) and the new business-trade zone, Sava's amphitheater. The traffic aspect is of great importance because the entire organization of the Belgrade Central Zone (Fig. 7.1) is designed to enable the creation of well-organized pedestrian spaces, ambiences, and streets and paths, both within and outside the identified central cores, on the riverbanks, over the river and within residential areas. This form of city communication, served by traffic systems and rich urban content and activities, is aimed at "liberating a city man for an old, yet new type of movement in the city" (Petrović and Ferenček 1977).

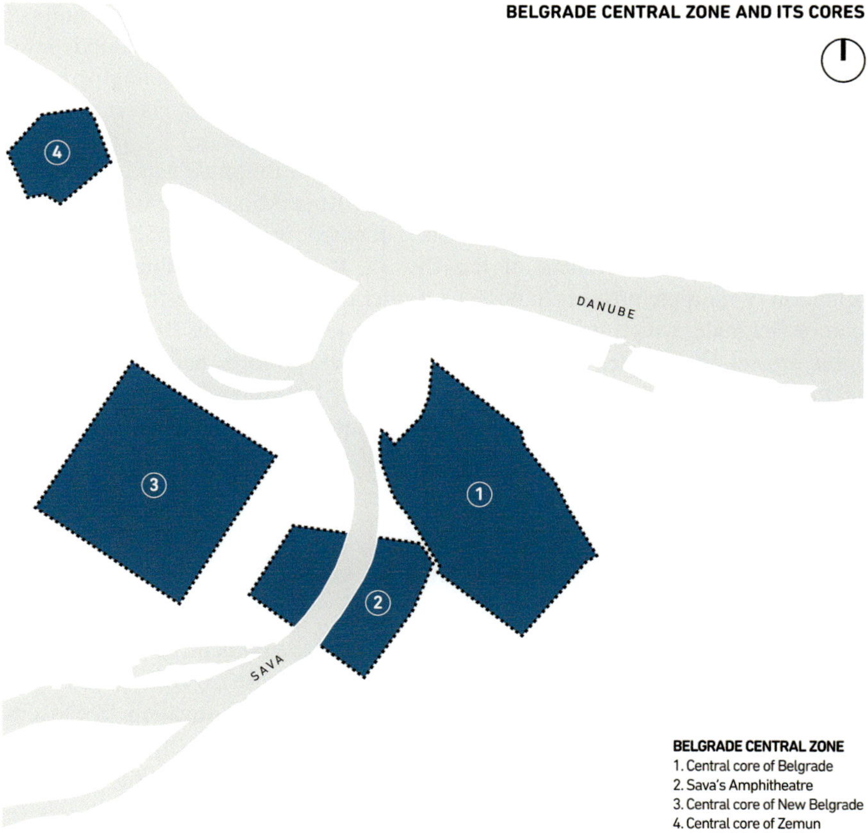

Fig. 7.1 Belgrade Central zone and its cores. Vukmirovic 2019

Following the spatial concept of the four main areas within Belgrade's central zone from the 70s (Solujić 2018), the process continued to the early years of the twenty-first century. Some of the works carried out in the last few years, beginning in 2015, show the continuity of these ideas (Fig. 7.2). In the old core of Belgrade, they include the extension of Knez Mihailova Street and its connection with the surrounding areas in order to make direct connections with the riverbanks; the introduction of the new pedestrian zone in the Savamala district; the development of the Skadarlija district; and improvement of local public spaces based on the principles of the green hierarchy in transport. In the core of the Sava amphitheater, within the ongoing implementation of the Belgrade Waterfront project, work is being carried out on the arrangement of the promenade on the left bank of the Sava River, called Sava Promenade, and Savski square, one of the most important squares in the city. The core of New Belgrade has had some spontaneous interventions that could be characterized as those that stimulate pedestrian movement

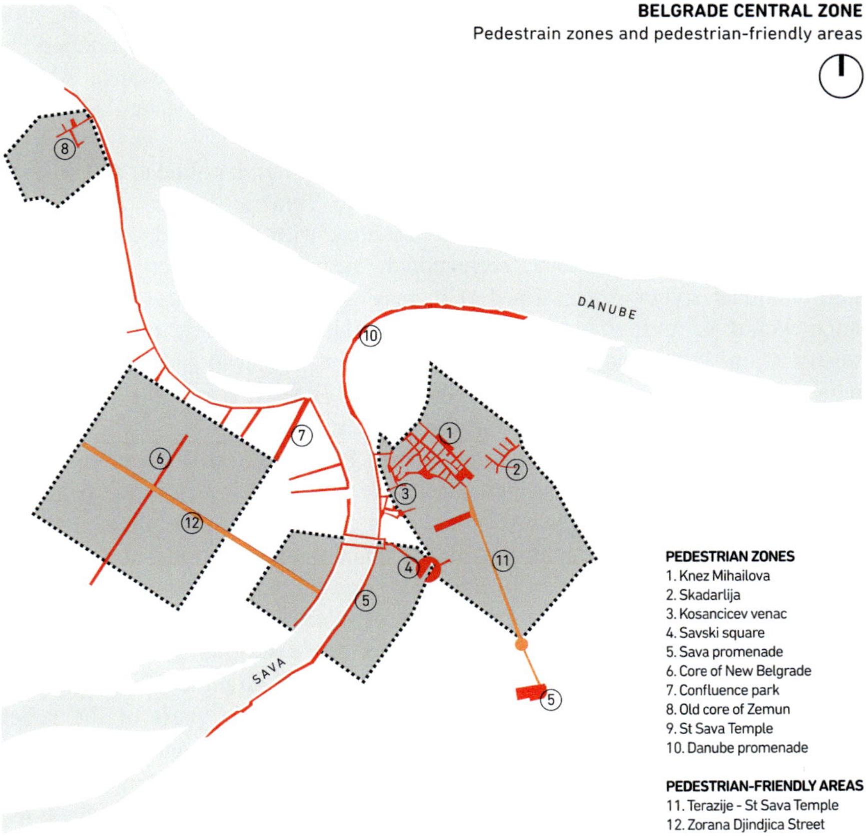

Fig. 7.2 Pedestrian zones and pedestrian-friendly areas of the Belgrade Central Zone. Vukmirovic 2019

along Zorana Djindjića Boulevard (Milaković and Vukmirović 2013) in Block 21, namely, the planned development of the pedestrian paths in block 26 and renewal of Friendship park (Park prijateljstva). These will improve the pedestrian connections between the core of New Belgrade and the Danube riverbank. The old core of Zemun, as defined by the Detailed Urban Plan from 2003, is slowly getting its planned look through the introduction of a pedestrian zone that connects Glavna Street with the Danube promenade.

Pedestrian areas are located in the most valuable parts of the city center and connect the most important city places and areas. The proposed spatial concept from 1976 stated that pedestrian zones should cover the spaces of Knez Mihailova Street, Republic Square, Nikola Pašić Square, Terazije Terrace, Slavija Square, the central area of the Sava amphitheater, the zone of the new congress center, and the centers of New Belgrade and Zemun. The other parts of the central zone should integrate pedestrian and motor traffic in order to avoid mutual negative effects.

An expansion of the center of gravity in the 70s was planned, with new interventions toward the Sava amphitheater, and the opposite area of the left bank planned for state representative contents such as a congress center. The most important linear urban element, called "the backbone" of the city, was defined by the old core of Zemun, Lenin Boulevard (Mihajlo Pupin Boulevard), AVNOJ Boulevard (Bulevar Zoran Djindjić Boulevard), Sava amphitheater, Nemanjina Street, and Slavija Square.

In the 70s, the Sava amphitheater was seen as the cultural, political, and business center of Belgrade, with integral connections to Terazije Terrace and Terazije Square. Improvements were planned for the area, with the introduction of commercial, cultural, entertainment, recreational, business, and other content. This multifunctional solution was aimed at increasing the attractiveness of the space, which would be further enhanced by the introduction of new types of public transportation. Special attention was also given to the renewal and care of specific ambient units—Kosančićev Venac, Skadarlija, Hotel Bristol, the area along Makedonska and Nušićeva Street and the Old core of Zemun.

The Regulatory Plan for the central zone of Belgrade from 1978 predicted the establishment of pedestrian zones and a network of pedestrian streets and paths, but by 1988 only part of the original plan had been completed. The implementation of elements of this document continued in the contemporary period, more precisely after 2000, through the adoption of new plans and initiatives, but essentially, they did not deviate much from the basic concept from the 70s. In addition to the expansion of the area of Knez Mihailova Street as a pedestrian zone, the General Plan of Belgrade 2021 envisages the spread of pedestrian corridors between Kalemegdan and Slavija along Kralja Milana Street. Also, parts of the streets toward Kosančićev Venac, from Slavija to St. Sava's Temple and the area from Nikola Pašić Square to Bezistan and the interior of the block are covered by the plan. It includes the reconstruction of Kralja Milana Street with Terazije, which was completed in 2002. The emphasis on the significance of pedestrian zones and the pedestrian network in the city center has also been noted in strategic documents such as the City of Belgrade Development Strategy, the Tourism Development

Strategy of the City of Belgrade, and the Strategy for Sustainable Development of the Stari Grad Municipality.

The most definite proposal for the expansion of the pedestrian network is given in the Tourism Development Strategy of the City of Belgrade. The strategy envisages the improvement of accessibility, with priority given to the most attractive zones, open spaces, roads, and accommodation that are well connected by public and/or specialized transport, and pedestrian and bicycle movement (Institute of Economic Sciences 2008). The research that preceded the development of the strategy found that the least work had been done on the design of new pedestrian zones and routes that would connect tourist areas on foot. Based on these results, and in accordance with the suggestions and solutions of the GUP of Belgrade for 2021 and the Regional Spatial Plan of the Administrative Area of Belgrade, a proposal was made for the fastest, cheapest, and most efficient solution (Institute of Economic Sciences 2008). The proposal includes the formation of a large tourist pedestrian zone, with the Sava and Danube as aqua boulevards, which would also be the backbone of the Belgrade tourist product called "City break". The zone would include parts of Belgrade's central zone and a number of urban units: Varoš u Šancu, Terazije, Slavija Square, Svetosavski Plateau, Great War Island, the old core of Zemun, parts of the original New Belgrade, and the core of New Belgrade (Institute of Economic Sciences 2008, 227). Selected urban units would be connected by forming: a walkway from Gardoš, through Zemun, the Hotel Yugoslavia and the Palace of the Federation (Palace of Serbia); a pedestrian bridge to the unique ecological educational complex of the Great War Island; and other appropriate connections through the river, Donji to Gornji grad, through Kalemegdan, Knez Mihailova Street, Kralja Milana Street, and Slavija to Svetosavski Plateau.

7.2 The Old Core of Belgrade

The core of old Belgrade is known as the center; it has the oldest and most valuable monument and heritage ambient of the city. It contains material testimonies of the duration of Belgrade from Roman times to the present day (Fig. 7.3). Perceived

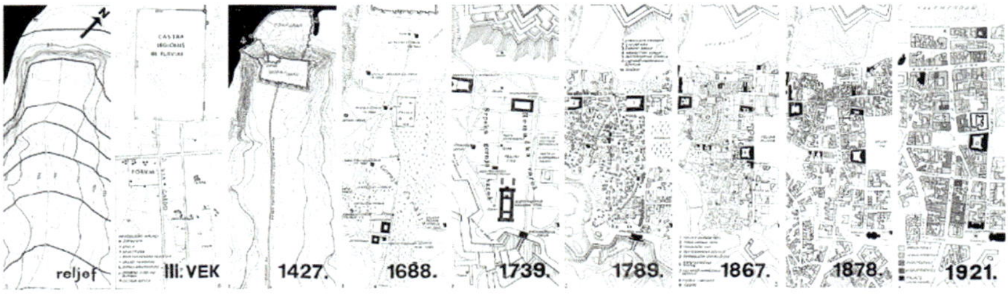

Fig. 7.3 Development of Knez Mihailova. Archive Belgrade Urban Planning Institute

from the aspect of the pedestrian network, this area includes the pedestrian zone of Knez Mihailova Street, whose expansion aims to meet the objectives of the plan for the central zone of Belgrade from the 70s.

The year 1867 marked a turning point in the history of the modern urban development of Belgrade. It was the year in which Emilijan Josimović, the founder of modern Serbian urban planning and the first Serbian urbanist, completed his inventive urban plan of modern European Belgrade, which formed the basis of a survey that he started as an individual enthusiastic researcher. Today, we refer to his yearlong work as the first modern urban plan of Belgrade, but it was actually Josimović's book on the reconstruction of Belgrade within the Moat that made the biggest impact, in which he explained his proposal for new urban regulations with a plan in the scale of 1:3000. His work was commissioned, in order to transform the oriental town into a European city (Terzić 1997; Medaković 1997; Maksimović 1967).

According to Blagojević, "Josimović was a contemporary to the famous European urbanists Haussmann and Cerdà, and his plan followed soon after their famous plans for reconstruction of Paris (1853–71), and Barcelona (1859–70). … His plan demonstrates the universality of the European planning paradigm of the period, and more precisely of the ring principle which was employed in Vienna, yet it clearly brings out the specific character of Belgrade into its proposals for reconstruction" (Blagojević 2009). Emilijan Josimović's work is of significant historical importance. The proposals and ideas from his book included, for example, connections between the fortress and the city. He proposed narrow roads through three gates, as three additional passages into the new town. Most crucial for the life of the city was definitely his proposed exit from Delijska (Knez Mihailova Street) to Terazije (Fig. 7.3). This new link emphasized the importance of Knez Mihailova Street, transforming it into a new trade center.

New interest in Knez Mihailova Street was aroused in the late 60s and beginning of the 70s with the draft of a new General Urban Plan of Belgrade. The new General urban plan of Belgrade 1972 introduced a new, modern concept of transportation planning in the city, based on a massive public transport system—the metro, as well as planning of the pedestrian zones. This can be considered as a partial continuation of a European trend of the time, having in mind that the first pedestrian streets in Europe were constructed between 1972 and 1978 (Vukmirović 2013).

The entire complex of Knez Mihailova Street (Fig. 7.4) was based on three strong urban roads (Vasina, Carice Milice, and Pariska streets), as well as a crossroads (Kralja Petra and Vuka Karadžića streets). Knez Mihailova and a series of adjacent streets were transformed into pedestrian streets as a result of the Decision on Regulating Traffic in the Central Zone in 1975 (Banović and Jakšić 1978). Having in mind all the above, the process of protection and arrangement of Knez Mihailova street formally began on December 11, 1975. Belgrade got its first pedestrian street in 1988 on the basis of the regulatory plan for the central zone of Belgrade from 1977 (Fig. 7.5).

The fundamental reconstruction of Knez Mihailova Street and part of Republic Square was carried out according to the project by architect Branislav Jovin. Work

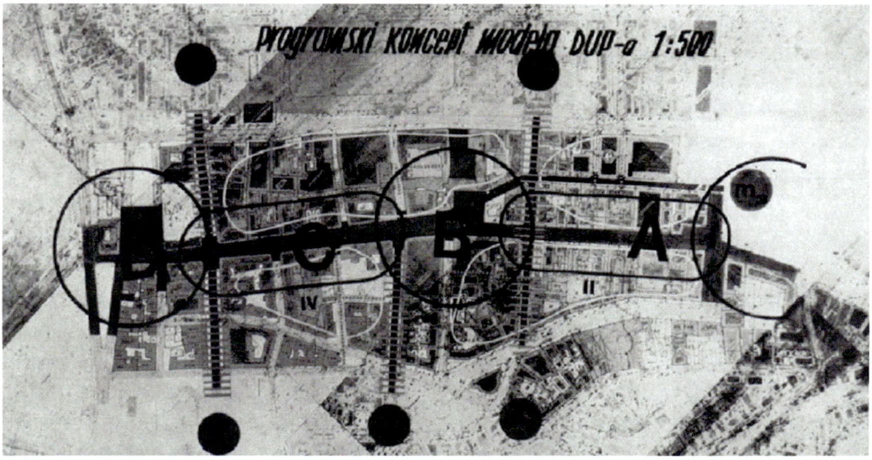

Fig. 7.4 Detail urban plan of Knez Mihailova Area—Program concept. Archive Belgrade Urban Planning Institute

Fig. 7.5 Knez Mihailova Street—before and after its transformation into a pedestrian zone. Archive: Belgrade Urban Planning Institute

started in May, and the street opened to pedestrians on October 20, 1988. According to Ferenčak, converting Knez Mihailova Street into a pedestrian zone was one of the few large urban themes that were brilliantly realized, bearing in mind that Belgrade had not developed enough in the areas of planning and continuity (Ferenčak 2008).

The Regulatory Plan for the Central Zone of Belgrade from 1977 envisaged the formation of a pedestrian zone, i.e., a network of pedestrian streets, but by 1988 only part of the plan had been realized (Vukmirović 2013). Expansion of Knez Mihailova Street pedestrian zone continued in 2008, after which nine more streets were completely or partially pedestrianized (according to the Plan from 1977). By mid-2010, most of the planned interventions of this kind were completed.

Following this and concluding with the reconstruction of Kralja Petra Street from Knez Mihailova to Cara Lazara Street, the pedestrian zone encompassed Čika Ljubina, Zmaj Jovina, Marka Leka, Laze Paču, Djure Jakšića, 1300 Kaplara, Nikole Spasića, part of Obilićev Venac, and part of Vuka Karadžića streets. A special regime for pedestrian traffic was defined in 27 streets, and apart from the pedestrian zone in the area of Knez Mihailova Street, a pedestrian zone was mapped in the area of Skadarska Street. This includes Emilijana Josimovića, Zetska, and part of Simina streets, and the zone belongs to the bohemian quarter known as Skadarlija.

This trend continued, but significantly more ambitiously, in 2015 with the IME project: Identity_Mobility_Environment (Folić and Vukmirović 2015). The project envisages the expansion of the Knez Mihailova Street pedestrian zone and includes the full implementation of the Plan of 1977, which would make a complete entity with Knez Mihailova Street as the focus and result in continued expansion of the pedestrian streets toward the bank of the Sava on one side and toward Dorćol, that is, the bank of the Danube on the other side. Extending the pedestrian zone of Knez Mihailova Street includes the implementation of multiple projects. In this way, the city will develop "a green network" of pedestrian spaces in its center, which would interconnect the existing public spaces and spatial units of a broader significance for its urban identity. The aim of this is to reduce the use of private cars for transport and to create a greener, healthier, and more pleasant living space. The project will be realized in four phases (Fig. 7.6).

The first phase of the project covers public space with a total area of 35,272 m^2. A small part of this area is now in the implementation phase (started in July 2015) and will be reconstructed on the basis of the initial project for the pedestrian zone of Knez Mihailova Street, designed by architect Branislav Jovin during the 80s. Other parts of this area were designed and completed by the end of 2018. The second

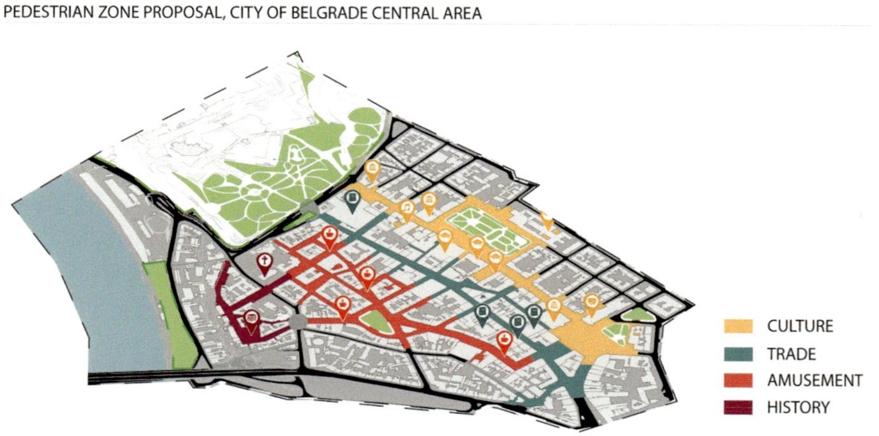

Fig. 7.6 Concept of the extension of the pedestrian zone around Knez Mihailova Street. *Source* Ana Delipara, Office of the Director of urban planning of the city of Belgrade

phase was planned be carried out simultaneously with the first phase. It includes an area of 10,149 m^2 located in Kosančićev Venac, which is under cultural heritage protection as an ambience ensemble. It was designed following the strict conditions of cultural heritage protection and realized in June 2017. The third phase of the project includes the street network in the Savamala district with a total area of 4,405 m^2. It will be carried out as part of Belgrade Waterfront project. Finally, the fourth phase envisages the reconstruction of streets and public spaces along Vasina and Uzun Mirkova Streets with a total area of 39,088 m^2. This area should be designed and completed by the end of 2019.

It can be seen that the phases are determined on the basis of the character and status of the territory, which will in the future result in the creation of four distinctive wholes: a hospitality zone (Topličin Venac with associated streets), a historical zone (Kosancicev Venac), a cultural zone (the path along Vasina and Uzun Mirkova Streets with Student square and the Plateau in front of the Faculty of Philosophy), and a trade zone (Knez Mihailova Street).

For the purposes of verification and further elaboration of this concept, the city authorities produced two documents—the Spatial–ecological valorization of the future pedestrian zone of Belgrade (Glumac et al. 2015) and the Feasibility study for enlargement of the Knez Mihailova Street pedestrian zone (CEP 2016). According to the valorization (Glumac et al. 2015), the extension of the Knez Mihailova Street pedestrian zone is the "most effective way of achieving environmental protection from the consequences of intense traffic, primarily for protection against noise and air pollution with the aim to reduce its level of the central zone of the city." In addition to this, the study (CEP 2016) provided different dimensions and variants for implementing the project. The study included various analyses and produced three possible alternative solutions. The expert team used different methodologies based on a literature review, with good examples of different pedestrian zones as case studies, as well as direct surveys of residents and car owners, models of intensity of transport, etc. The study also defined different goals such as general conditions of bicycle transport in the center of the city, logistics inside the area of the pedestrian zone, smart functioning of the garbage disposal system, organization of public events, development of the GIS database, and the functioning of taxis and tourist transport. In this way, both documents reflect a general positive attitude with regard to this project and initiative.

7.2.1 Development of Public Spaces Within the Old Core of Belgrade

The first segment of the project—the reconstruction of Vuka Karadžića Street—was finished and officially opened in December 2015. It was done in accordance with the project by architect Branislav Jovin, dated from the first phase of the reconstruction of the street (1986). The second segment of the project—the

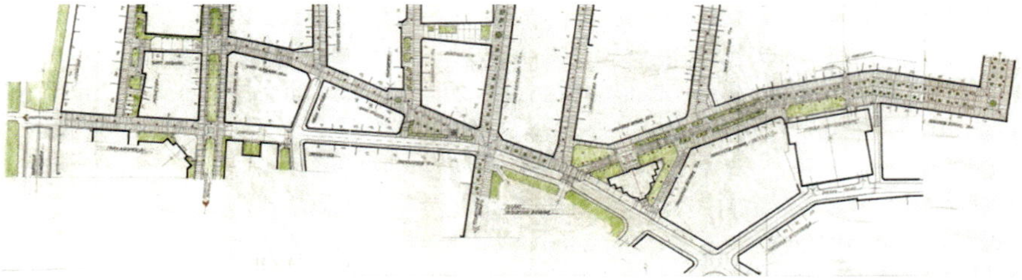

Fig. 7.7 Concept of the reconstruction of Obilićev Venac and associated streets designed by Branislav Jovin and Siniša Temerinski. Archive: Office of the Director of urban planning of the city of Belgrade

reconstruction of the Red Rooster (Crveni Petao) Plateau—was finished and officially opened in April 2016, while the third segment—Zmaj Jovina Street—was reconstructed in August 2016, also in accordance with the project by Branislav Jovin. In May 2016, the same team gave the final proposals for the area of Oblićev Venac (Fig. 7.7). The reconstruction of this area began on September 3, 2016.

The second phase of the pedestrian zone extension covers the areas on the sides of the existing zone. It will be developed in the form of the reconstruction and improvement of individual open public spaces and paths in order to establish a connection with Belgrade's riverbanks (on the Sava and Danube). The team that will deliver the public space design proposals is led by Professor Boris Podrecca and the interventions include four segments in the immediate vicinity of the Knez Mihailova pedestrian zone.

The first segment covers the path along Vasina and Uzun Mirkova streets with some focal points—public spaces that will be formed on Student Square and on the plateau in front of the Faculty of Philosophy (Fig. 7.8). This pedestrian path will directly connect these public spaces with the area of Belgrade Fortress, Kalemegdan Park, and Republic Square. The main idea is seen in the connection between the two open public spaces, the square and the park, using the organic structure of a stone pavement. In this way, an urban entity is formed which is visually connected with the park, and which at the same time offers features and contents characteristic

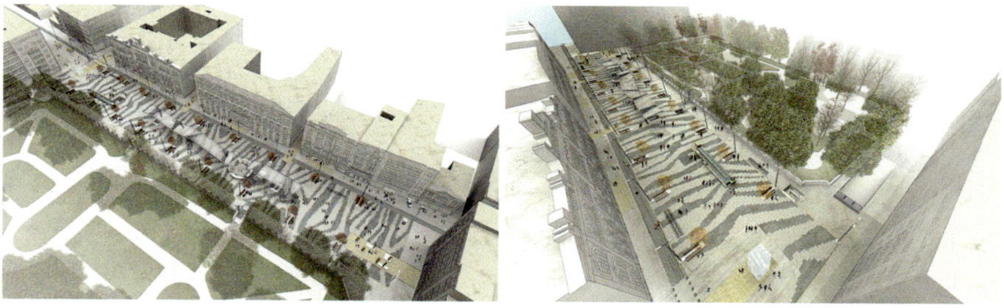

Fig. 7.8 Concept design of Student Square. Author Boris Podrecca. Archive: Office of the Director of urban planning of the city of Belgrade

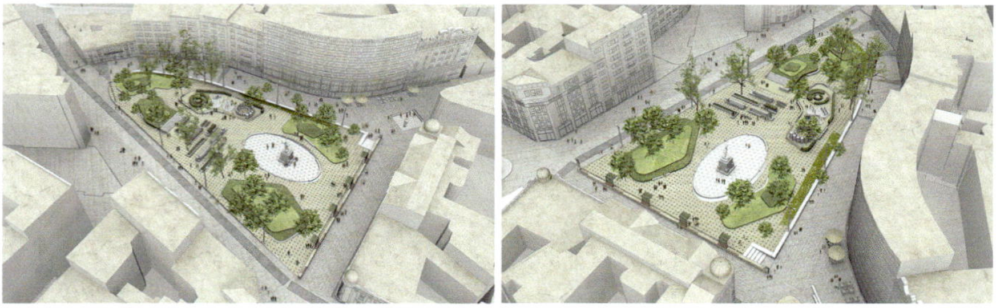

Fig. 7.9 Concept design of Topličin Venac. Author Boris Podrecca. Archive: Office of the Director of urban planning of the city of Belgrade

of a city square, and includes the outdoor lobby of important buildings such as the University of Belgrade, the Kolarčeva foundation, the Faculty of Philology, and the Ethnographic museum.

Segment two belongs to Topličin Venac and its associated streets. The main public space in this area will be Vojvoda Vuk Park with a specially designed area for children. Topličin Venac Street, together with Vuka Karadžića Street, will form a link that will make a direct pedestrian connection with Kosančićev Venac, which is seen as the historical core of both Belgrade and the modern Serbian state (Fig. 7.9). Kosančićev Venac was renewed during 2017, following the strict conditions for protecting architectural heritage, by which the place received its recognizable character, attracting both old and new visitors (Fig. 7.10).

The site of the ruined National Library on Kosančićev Venac is seen as an historical memory labyrinth and it belongs to segment three. The concept was designed by Boris Podrecca, who had the goal of creating a meeting and livable place, which also attracted visitors with its rich history (Fig. 7.11).

The fourth segment is defined by the path along Sremska and Maršala Birjuzova Street, one of the busiest pedestrian connections between Knez Mihailova Street

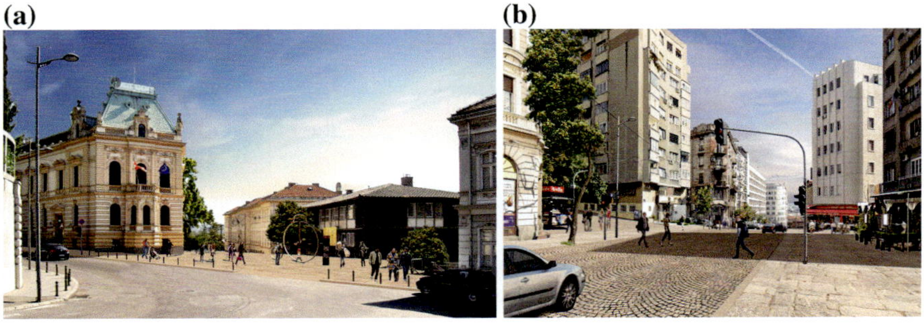

Fig. 7.10 Main entrances to the area of Kosančićev Venac: **a** from Pariska Street and **b** from Topličin Venac. *Source* Mašan Petrović 2017, Office of the Director of urban planning of the city of Belgrade

Fig. 7.11 Memory labyrinth at Kosančićev Venac. Author Boris Podrecca. Archive: Office of the Director of urban planning of the city of Belgrade

and the Zeleni Venac area. Part of this segment was completed in 2017 covering Maršala Birjuzova Street (from Pop Lukina to Sremska Street) and Kosmajski Passage, which connects Maršala Birjuzova Street with the newly pedestrianized area of Oblićev Venac.

Since the metro has not been built, the functioning of the new pedestrian zone will be possible by introducing a transportation ring for public transport, as well as service and automobile use (most of it will belong to a 30 km/h zone). On the other hand, access gates to the pedestrian zone (for public transport and private cars) are planned for the locations of public transport stops and the planned locations for the new public underground parking. For elderly and disabled people, as well as deliveries, new electric cars that can be moved in circles in the zone will be introduced.

In line with this project is the idea of the renovation and arrangement of the three main city squares: Republic Square, Terazije Square, and Nikola Pašić Square. An international urban–architectural competition was announced in 2015 in order to obtain an integral vision for the transformation of these public spaces and their surroundings.

The first prize was given to the proposal by a team led by Zoran Dmitrović and Zorica Savičić, who proposed a comprehensive solution for the entire competition area designed to suit pedestrians (Fig. 7.12). Implementation of the first phase of this project started in 2018.

Most of the works on the extension of the Knez Mihailova pedestrian zone are still in progress, and their completion is expected by the end of 2020, after which it will really be possible to measure the success of the ideas that were defined almost 50 years ago. These interventions include the reconstruction of Karadjordjeva Street and the Sava riverbank that will be connected with Belgrade Liner Park. This will stretch from the area of Belgrade Waterfront to Ada Huja, i.e., along almost the entire right bank of the Sava and Danube in the central zone. This is presented in more detail in Chap. 4, which deals with the development of the riverfront.

Fig. 7.12 Winning proposal for the Three Squares competition. Authors: Zoran Dmitrović, Zorica Savičić, Nenad Blagojević, Ana Branković, Zoran Damjanović, Natalija Živulović, Marina Ilić, Vladimir Lukić, Ivana Pohulek, Miloš Radojević, Marija Radojlović, and Djordjije Čizmović. Archive: Office of the Director of urban planning of the city of Belgrade

7.3 The Core of the Sava Amphitheater

The plan for the Belgrade central zone was defined by a network of pedestrian spaces and routes within the four cores, as well as direct connections between them. If this idea is fully realized, Belgrade will have one of the largest pedestrian zones in the world, and the concept itself could be characterized as extremely advanced, bearing in mind the period when it was created.

Direct connection between the old core of Belgrade and the Sava amphitheater is planned along a line that will be formed by building Terazije Terrace (Fig. 7.13), which is one of the ideas defined at the beginning of the twentieth century by French architect Alban Chambon. He foresaw the transformation of the center of Belgrade by establishing the main streets and squares, along which he defined the locations of the complexes and facilities of the main state institutions. One of these was Terazije Terrace. Following a similar concept of cascading terraces and gardens, another solution was proposed in 1921 by Austrian architects Emil Hoppe and

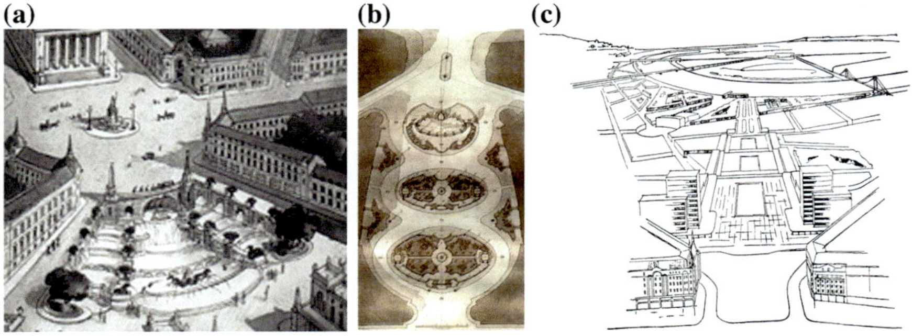

Fig. 7.13 Terazije terrace: **a** Chambon (1912), **b** Hoppe and Schonthal (1921) and **c** Dobrović (1929/30). Archive: Vukmirović

Fig. 7.14 Terazije Terrace. Author: ARCVS, September 2016. Archive: Office of the Director of urban planning of the city of Belgrade

Otto Schönthal (Vuksanović Macura 2013). The idea of a direct connection between Terazije and the banks of the Sava River was proposed by architect Nikola Dobrović in 1929/1930. The path was shaped in a modernist style, forming a cascading park, surrounded by residential and commercial buildings.

Terazije Terrace has been an urban design topic for decades and has captured the attention of architects and urbanists, with two urban architectural competitions being announced in search of the best solution. The first one was in 1991, when Professors Slobodan Mica Rajević and Zoran Nikezić won the first prize. The project has long been only an idea on paper, with a partial realization in 2010 that included one building and its access from the park side by Professor Vasilije Milunović. The authors of the winning design were members of Milunović's team on this project. The second competition for this location was in 2007. Two project bureaus were awarded first place in this competition: RE:ACT and ARCVS. The pedestrian connections in the projects along Terazije Terrace are in the form of large steps with a slight incline, with one of the designs proposing a cascading fountain along the steps. After years of silence on the theme of Terazije Terrace, the ARCVS proposal (Fig. 7.14) was updated with some modifications and its completion is expected in the next 5 years.

From the perspective of pedestrian networks, Sava amphitheater is important for three reasons: (1) It is seen as one of the cores of the Belgrade central zone, (2) it includes a significant part of the Sava waterfront (Sava promenade), and (3) it is

Fig. 7.15 Sava Promenade. Archive: Office of the Director of Urban planning of the City of Belgrade

seen as a place for two important pedestrian areas/spaces, one within the Savamala area and the other in Savski Square.

Sava Promenade is conceived as a collective pedestrian place where "residents can delight in the peaceful flow of the river while they enjoy food, music, and culture" (Aquino and Hung 2016) along the Sava River. The project is divided into three segments including the riverfront zone, the zone of Hercegovačka Street, and the promenade zone. The 1.8-km-long path will connect the area of Savamala with Ada Ciganlija, one of the most visited recreational areas in the city. The design of the Promenade, an integral part of Belgrade Waterfront project (Fig. 7.15), was created by the SWA group, a landscape design studio based in Los Angeles led by Gerdo Aquino. Work on the project started in 2015. It is still an ongoing project, completed in phases, but almost half of the planned interventions have already been completed with the space being widely used in the expected way.

The arrangement of two areas in close proximity to Sava Promenade is planned. The first space is directly connected with the Sava Promenade through Hercegovačka Street and it includes the pedestrian zone in Savamala. It is formed by Travnička, Mostarska, and Kraljevica Marka streets, including Travnički Park in front of the Geozavod building. This area of Savamala is recognized as Belgrade's Creative District and is a quality nightlife area. Part of this area was the subject of an international competition for Savski Square, while the project for Kraljevića Marka Street (Fig. 7.16b), as a connection between Zeleni Venac and Savamala,

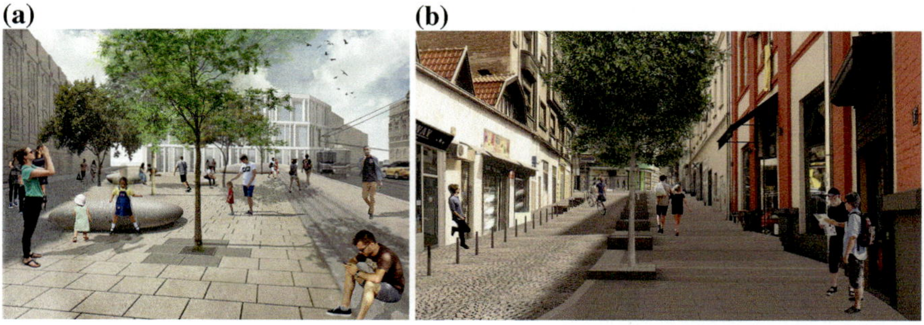

Fig. 7.16 Savamala pedestrian zone. **a** Competition design for Travnički Park, authors: Vujović, Vukmirović, Petković, and Ilić 2019 and **b** design proposal for Kraljevića Marka Street. Authors: CreativeBGD

was designed by the CreativeBGD team in 2017 and it is waiting to be implemented.

An international urban–architectural competition for Savski Square was announced in November 2018. As already mentioned, as a subject, it included two areas connected by Karadjordjeva Street: the area of Savamala with the focus on Travnički Park (Fig. 7.16a) and the area of Savski Square, one of the most important public spaces in the city. Savski Square is situated in front of the building of the old railway station, which is planned for renewal and will become a museum of Serbian history. The space is an intersection of several important city routes such as Karadjordjeva, Savska, and Nemanjina streets, which connect it with other important parts of the city.

Savski Square is perceived as a meeting place, an outdoor lobby of the Museum of Serbian history and the main entrance into the Sava amphitheater area. It will be organized as a pedestrian area, surrounded by concentrated streets with tram, bus, and cycling lanes. A monument to Stefan Nemanja, the founder of the Serbian state, is planned for this square, making it even more significant in symbolic terms. The international competition resulted in over 25 solutions. The Spanish team, Fenwick Iribarren Architects, was selected as having the best solution and awarded the first prize (Fig. 7.17).

They proposed a circular pedestrian area surrounded by a green ring, after which the project was named. The plan raises the space in front of the museum building in relation to rest of the square, simulating a stage. The stage is seen as a place for different events and activities, as well as the place where the monument will stand. The winning proposal will be the basis for the development of the urban plan and detailed urban design project, after which its implementation will begin. Savski Square will receive its facelift in 2020.

Fig. 7.17 Savski Square proposal designed by Fenwick Iribarren Architects. Archive: Office of the Director of Urban planning of the City of Belgrade

7.4 The Core of New Belgrade

The center of New Belgrade is recognized as its core. It consists of nine urban blocks, planned in accordance with the principles of the modern movement. Three of them are 400 × 400 m and the other six are 400 × 600 m, together forming a rectangular urban area. It has potential to become a part of Belgrade's pedestrian network, since the main axes of pedestrian movement are the path that goes along Zoran Djindjić Boulevard and the line that connects the government building of Serbia with the Belgrade Arena sports hall. As mentioned earlier, Zoran Djindjić Boulevard is a busy traffic road, and a planned extension will directly connect it with the core of New Belgrade and Sava Amphitheater, as defined by the Belgrade central zone concept (Fig). This idea was described in greater detail in the proposal for the Sava amphitheater congress center by Professor Miloš Perović in 1975.

The detailed plan for Blocks 25 and 26 was developed based on the winning urban design proposal for this location. It primarily refers to land in Block 26. The main features of the concept are freestanding objects positioned along the perimeter of the block, shaping the free central space of a square with a park arrangement. The central square integrates the free area around the church complex, creating a unique public space available to all users (Fig. 7.18). Within the block, the construction of internal overground roads will not be permitted, and an automobile approach will be made possible by the construction of underground access roads.

This line directly connects New Belgrade's center with the area of the Danube waterfront through Block 13 and the area of Confluence Park on Blocks 14 and 15, together with its parallel paths: Art Boulevard (Bulevar umetnosti), Španskih

Fig. 7.18 Blocks 25 and 26 and their connection with the Confluence Park. Authors: Miljković D Mitrović J. Archive: Office of the Director of Urban planning of the City of Belgrade

boraca Street, Antifašističke Borbe Street, and Milentija Popovića Street. Block 13 is one of the largest blocks (almost 400 × 1,600 m) in New Belgrade and it is where the famous Serbia palace (Ex-federation Palace or SIV Palace, SIV, FEC—federal executive council) is situated, between Mihajlo Pupin and Nikola Tesla Boulevards. It was the subject of the international urban architectural competition for the Philharmonic building which, after much discussion about its future location, finally found its place here (more about this project and the competition is written in Chap. 3).

Blocks 14 and 15, well known as Friendship Park, from the pedestrian perspective were the subject of an analysis and vision proposal by Gehl Architects at the beginning of 2017. The main findings from the analysis were that it was difficult to enter, with low-quality routes and few variations in the experience. On the basis of these findings and from the standpoint of place identity, Gehl (2017) proposed the concept "from a green void with a weak identity to clear identity as a people's park." In accordance with this, they defined three strategic paths in order to attract and generate life in the area of Friendship Park (Fig. 7.19). The first path is related to the problem of entry into the park, whereby the park should be connected with the city. As a second path, a "fine-grained network of paths with a clear hierarchy" was proposed. Finally, a plan was made to "maximize existing amenities and invite people to engage and interact" as a third path.

From the aspect of extending the pedestrian network, the most interesting part of the project is the introduction of seven thematic routes that will make direct connections between Block 13 and the Danube riverbank. Based on the context of the location, thematic routes include the following titles: move, nature, Serbia, music, water, art, and science (Fig. 7.20). They are distributed in relation to the main character of that particular part of Friendship Park and its connection with its surroundings.

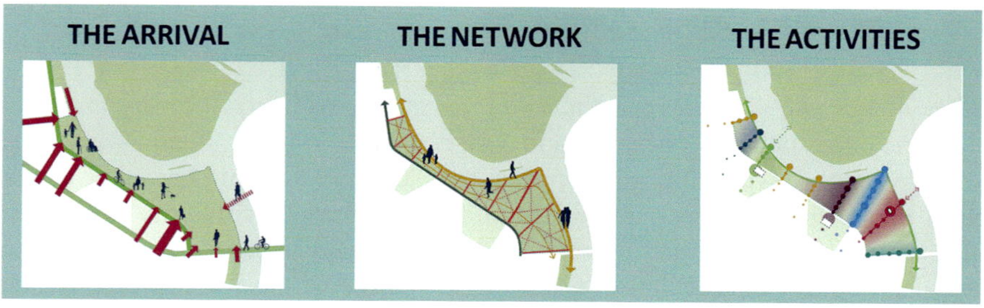

Fig. 7.19 Strategic paths: the arrival, the network, and the activities. Author: Gehl Architects. Archive: Office of the Director of Urban planning of the City of Belgrade

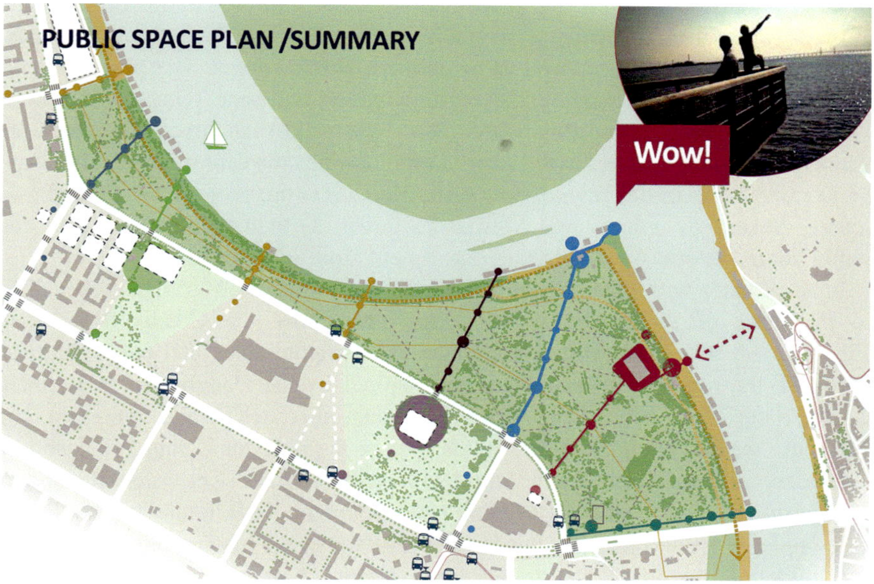

Fig. 7.20 Park Ušće—People's Park. Spatial distribution of the thematic routes. Author: Gehl Architects. Archive: Office of the Director of Urban planning of the City of Belgrade

Work on the project started in 2017 and one of the routes, the art route, is currently being constructed. This project is ongoing, and some parts of its implementation depend on other projects like the Philharmonic, Gondola, and the Natural Museum. At the same time, the completed part has shown the expected results, and this area is used more than it was before.

7.5 The Old Core of Zemun

Belgrade has two old cores that "make up its organic parts and are very important elements in its spatial organization and design" (Banović 1973). It gained its actual form during the eighteenth century, when the old core of Zemun became a central point in the social, economic, and cultural development of the city of Zemun. Because of its character and historical value, during 1979 the old core of Zemun was declared a spatial, cultural, and historical entity of great importance for the state. In accordance with its history, identity, and purpose, this area also required special design treatment and priority to be given to pedestrian movement and human size.

A detailed urban plan was adopted in 2003 for the area of the Old core of Zemun. The plan determined the scope and the concept of protection, arrangement, and renovation in order to preserve tradition, identity, and the cultural, historical and natural environment, as well as to improve public spaces. The plan also proposed retaining and expanding the existing pedestrian zone (Gospodska Street, Victory Square, Preka, and Piljarska Street). New pedestrian zones and streets are planned from Glavna Street to the river in order to open the center of Zemun toward the Danube bank. In this regard, new pedestrian zones are proposed in Omladinski Square, Masarikov square, and Mornarska Street, as well as Senski Square and within central Block A 27, where a department store is proposed.

In accordance with the plan, the pedestrian zone was improved in October 2017 on the basis of the public space design project by the CreativeBGD team (Fig. 7.21). It was made with a combination of granite slabs and cubes, and the motifs on the new pavement pattern are inspired by the Danube River. Its main characteristic is an uneven line where two materials touch, which represents the riverbank, the wave, and its flow. As defined in the plan for Zemun's old core, a pedestrian zone connects Glavna Street with Victory Quay (Kej Oslobodjenja).

The initiatives to extend the existing pedestrian zone in Belgrade's central zone are completely in line with the initiatives and plans of the leading cities in Europe. The guidelines proposed by the European Union in relation to the Sustainable Urban Mobility Plan have proved to be feasible and able to achieve the expected results. Although it is considered ambitious and its implementation will take several years, the project to extend the existing pedestrian zone in Belgrade has real potential. It has been positively evaluated in different studies, and it represents the continuity as well as the development of the core planning ideas and visions. However, the expected functioning of pedestrian zones and implementation of the new transport hierarchy is possible only if the entire policy of urban and transport planning in the city is based on the principles of the green hierarchy and human scale. This can complicate the particular situation in Belgrade, because the city still needs an independent massive transport reform like the one planned in the 70s, but the question is: what will be its final form?

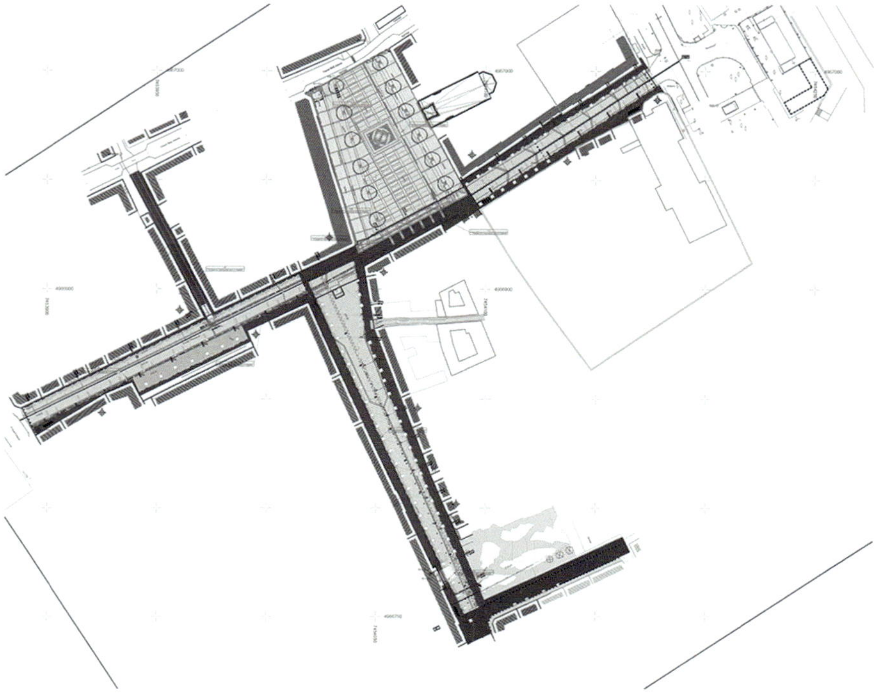

Fig. 7.21 Zemun pedestrian zone extension proposal. Author: CreativeBGD. Archive: Office of the Director of Urban planning of the City of Belgrade

References

Airparif (2014) Circulation alternée le 17 mars 2014. Accessed 26 July 2016, from Airparif: http://www.airparif.asso.fr/actualite/detail/id/112 (Airparif. Alternate circulation 17th March 2016, Available online)

Aquino G, Hung Y (2016) "Belgrade Waterfront Project—Sava Promenade—9a Biennale 2016". https://www.arquitectes.cat/iframes/paisatge/fitxa/10074

Banović A (1973) Staro jezgro Zemuna - Detaljni urbanisticki plan bloka 27. Urban Beogr 24: 5–7

Banović A, Jakšić M (1978) Prikaz dosadašnjih faza rada na Detaljnom urbenističkom planu Knez Mihailove ulice Urbanizam Beograda X:4–41 (Banović A, Jakšić M (1978) An overview of the current phases of the work on the Detailed Urban Plan of Knez Mihailova Street X:4–41)

Blagojević LJ (2009) Urban regularisation of Belgrade, 1867: trace vs. erasure. SAJ Serb Arch J 1:27–44

CEP (2016) Studija izvodljivosti proširenja postojeće pešačke zone Knez Mihailove ulice. Studija izvodljivosti, CEP, Beograd (CEP (2016) Feasibility study of enlargement of pedestrian zone Knez Mihailova street, Feasibility study, CEP, Belgrade)

Department for Transport (2007) The manual for streets. Thomas Telford Publishing, London

Eric J (2015) 6 Big European cities with plans to go car-free. The CityLab. Available on Internet: http://www.citylab.com/cityfixer/2015/10/6-european-cities-with-plans-to-go-car-free/411439/

European Commission Environment Directorate-General (2004) Reclaiming city streets for people: chaos or quality of life?. Office for Official Publications of the European Community, Luxembourg

Ferencak M (2008) Miodrag Braca Ferenčak, poznati urbanista – „gde ćeš, sine, baš na arhitekturu? ". (E-Kapija, Interviewer) (Ferencek M (2008) Miodrag Braca Ferencek, famous urbanist—"are you sure, my son, to study architecture?")

Folic M, Vukmirovic M (2015) Projekat IME: Identitet_Mobilnost_Ekologija Grada Beograda. Beograd. Grad Beograd - Sluzba za informisanje (Folic M, Vukmirovic M (2015) Project IME: Identity_Mobility_Environment. City of Belgrade—Department for Communication, Belgrade

Fouche G, Solsvik T (2015) Oslo aims to make city center car-free within four years. Yahoo News. https://www.yahoo.com/news/oslo-aims-city-center-car-free-within-four-124930728.html?ref=gs

Gehl Architects (2017) Ušće park—people's park. Gehl Architects, Copenhagen (digital source)

Glumac S, Sterđević B, Košpić S (2015) Prostorno-ekološka studija buduće pešačke zone Beograda. Prostorno ekološka valorizacija, Beograd. Grad Beograd – Sekretarijat za zaštitu životne sredine (Glumac S Sterđević, B Košpić S (2015) Spatial-ecological study of the future pedestrian zone of Belgrade. Spatial ecological valorisation, Belgrade. City of Belgrade—Secretariat for Environmental Protection)

Institute of Economic Sciences (2008) Strategija razvoja turizma grada Beograda. Beograd. Gradska uprava - Sekretarijat za privredu, Grad Beograd ((2008) City of Belgrade tourist development strategy)

Maksimovic B (1967) Emilijan Josimovic - prvi srpski urbanist. Beograd: IAUS (Maksimovic B (1967) Emilian Josimovic—the first Serbian urbanist. IAUS, Belgrade)

Medakovic D (1997) O Emilijanu Josimovicu (izabrani tekstovi). Beograd: Drustvo urbanista Srbije (Medakovic D (1997) About Emilijan Josimović (selected texts). The Association of Serbian Urbanists, Belgrade)

Milakovic M, Vukmirovic M (2013) Analysing the qualities of being on foot in modern cities: potentials of pedestrian environment in New Belgrade. RESPAG 2013. Belgrade. 22–25 May 2013. IAUS, Belgrade

O'Sullivan F (2013) Madrid's big plan to swear off cars. The CityLab. Available on Internet: http://www.citylab.com/commute/2013/12/madrids-big-plan-swear-cars/7744/

O'Sullivan F (2014) Europe's most congested city contemplates going car-free. The CityLab. Available on Internet: http://www.citylab.com/commute/2014/01/europes-most-congested-city-contemplates-going-car-free/8027/

Petrović V, Ferenčak M (1977) Centralna zona Beograda (Central Zone of Belgrade). Urbanizam Beograda IX:27–35

Solujić Z (2018) Planiranje i Prostorni Razvoj Komercijalnih Delatnosti i Gradskih Centara u Beogradu Posle Drugog Svetskog Rata in 70 Godina Urbanističkog Zavoda Beograda. Knjiga IV: Eseji, Beograd: Urbanistički zavod Beograda, pp 39–45 (Soplujic Z. Planning and spatial development of the commercial contents and the City Centres in Belgrade after the WWII in 70 years of Belgrade Urban Planning Institute. Book IV: Essays. Belgrade Urban Planning Institute, Belgrade, pp 39–45)

Terzić D (1997) Emilijan Josimovic, vizionar ulaska Beograda u urbanu civilzaciju. Beograd: Drustvo urbanista Srbije (Terzić D (1997) Emilian Josimovic, visionary entering Belgrade in Urbane civilization. The Association of Serbian Urbanists, Belgrade)

Vukmirović M (2013) Značaj i uloga mreže pešačkih prostora u generisanju kompetitivnog identiteta grada. Doktorska disertacija). Beograd: Univerzitet u Beogradu - Arhitektonski fakultet (Vukmirovic M. (2013) The role and the importance of pedestrian spaces in generating the competitive identity of the city. PhD Thesis, University of Belgrade—Faculty of Architecture, Belgrade

Vusanović Macura Z (2013) Emil Hope i Oto Šental: Projekat za Terazijsku terasu/Emil Hoppe and Otto Schönthal: urban design project for Terazije Terrace, Belgrade. Nasleđe XIV:155–169

Chapter 8
Belgrade's Development Visions and Aims

Milena Vukmirovic

Abstract The City of Belgrade is the capital of Serbia and one of the largest Southeast European cities. During the twentieth century, it endured many changes in its urban structure and experienced a significant population increase. Because of the Yugoslavian war crisis, the state suffered international isolation that resulted in the economic, social, and cultural decline of Belgrade. After 2000, the transition process began, which is still going on even today. Serbia is on its way to full EU membership but is not likely to enter the EU until after 2025. Considering the hard start to the twenty-first century and the challenges of the contemporary period, this chapter reveals current urban development and presents the ambitions and long-term visions for Belgrade.

During the last decade, Belgrade has been seen as the Capital of Southeastern Europe (Palgo 2011), a "distinctive, attractive and competitive city" (Vukmirović 2016), an "urban, sustainable, responsible, dynamic and developed European region" (RAREI and Palgo centar 2017), a smart city (Folić and Vukmirović 2016), and a "green, healthy, friendly and pleasant city" (Radojičić 2019). The general development goals have differed over the years, which has affected the clarity of the vision for its long-term development. Considering this, the chapter will present an overview of the visions and aims for the development of the City of Belgrade, defined within its actual planning and public policy documents.

2018 marked 70 years of the Belgrade Urban Planning Institute (Fig. 8.1). This is important to mention, because this institution has been a key instrument in the creation of the urban image of Belgrade and in the proposals for its future development. The establishment of Belgrade's Urban Planning Institute is linked to the first post-war years, which preceded a period of extensive, and perhaps the greatest, renewal, and construction in the former capital of Yugoslavia, the present capital of Serbia. The Urban Planning Institute presented the first draft of the general traffic solution and land purpose for the wider area of Belgrade to a scale of 1: 10,000, in 1948, including Belgrade, New Belgrade, and Zemun. This draft is considered to be the first urban document for the whole territory of Belgrade, which was widely discussed and publicly presented. Although never adopted as an official urban plan, the study made an overview of all the previous considerations and decisions on

© Springer Nature Switzerland AG 2020

B. Arandelovic and M. Vukmirovic, *Belgrade*, The Urban Book Series,
https://doi.org/10.1007/978-3-030-35070-3_8

Fig. 8.1 70 years of Belgrade Urban Planning Institute. *Source* Belgrade Urban Planning Institute 2018

important issues and concepts, and in practice it served as an inspiration and expert basis for all further research, which resulted in the adoption of the final elaboration of the General Urban Plan of Belgrade 1950 (Glogorijević and Graovac 2018).

The Belgrade Urban Planning Institute team, led by Miloš Somborski, Djordje Šuica, and Josif Najman, made the General Urban Plan of Belgrade, adopted on October 20, 1950. The most important topic in this plan was the design of New Belgrade, considered as a new city on the left bank of the Sava, unified with Belgrade and Zemun. The aim was for it to become the future political, economic, and administrative center of the state. Other elements of the 1950 general urban plan included planning administrative and cultural centers for the needs of all of the citizens in the city, the republic and the whole country; passenger, rail, and water transport as separate systems; the systematic afforestation of the entire area surrounding Belgrade; solutions for basic infrastructure systems; and moving rail traffic from the right bank of the Sava River. Although based on methodologically underdeveloped tools, the result of this plan was significant. Its implementation was enabled by a good intuitive model, great enthusiasm for the rapid renewal and construction of the capital, and strong state control (Glogorijević and Graovac 2018). However, its key deficiency was the lack of a strategic component and adequately defined strategic goals.

8.1 Visions Defined by the General Urban Plans of Belgrade

Belgrade experienced a key turning point in urban planning in the early 70s (Arandelovic et al. 2017). The adoption of the General Urban Plan of Belgrade for 2000 in 1972 marked "the beginning of the change in practice of urban planning in Yugoslavia, from a technical document to a strategic process" (Glogorijević and Graovac 2018). In a TV show called "Belgrade in the twenty-first Century" recorded in 1974 (Momcilović 1974), one of the executives who developed the plan, Aleksandar Djordjević, presented it as "a sort of social agreement on the future development of Belgrade." On the basis of extensive research, strategic goals were established by the urban plan, including directions for the expansion of the city, the realization of railway shafts and the relocation of the railway station, the lowering of the city to the rivers, the construction of a metro system, the location of new residential settlements, the resolution of the accumulated traffic problems, the reconstruction of the central zone of the city, the location of industry, the construction of recreational areas, etc. This document is important because it has generally determined the overall development of Belgrade, and its guidelines are still being implemented. On the other hand, the concepts implemented in this plan from the beginning of the 1970s were also modern and innovative in a wider context and were applied much later in strategic planning in Europe. However, a poor estimate of the economic situation, including both economic and recessionary crises, prevented the realization of most of the objectives defined by the plan. During the 90s, Belgrade experienced a major degradation, the resolution of which required a new plan.

Development discontinuity in almost all domains, a damaged image, and challenges dictated by global trends demanded a fast, but wise response and the establishment of an appropriate framework of strategies and plans for the future. The response began with the adoption of the Master Plan of Belgrade 2021, developed by the Urban Planning Institute of Belgrade and adopted in 2003. The Plan has a dual character—both strategic and operational. At the same time, it sets the general goals for the city's development and is "open enough to provide the opportunity to choose the path when a society opts for a development model" (GUP 2003). In terms of its operational character, it has been used to define the basis for the direct implementation of construction rules and to ensure the "rehabilitation of the city and urbanism" (GUP 2003). Several modifications since 2003 reflect the steps of maintenance of the urban plan under the Law on Planning and Construction and the changing needs of the Republic of Serbia and the city of Belgrade.

The plan focuses on "several large city projects, the possibilities for individual construction, the arrangement of settlements on the periphery of the city, the reduction of industrial and working zones in the center and the rationalization of transport and traffic solutions" (Glogorijević and Graovac 2018). The identity of the city is a significant topic, and the three branches of the natural corridors of the Danube and the Sava with War Island (Ratno ostrvo) stand out as important elements. Besides lowering Belgrade to its rivers, which is a topic that continues from the previous plans, certain locations stand out as being significant and strategic, namely, Sava amphitheater on the left and right bank of the Sava, Belgrade Fortress, the new economic zone along the highway, the new island and the recreational zone on the Danube, the railway station "Center" in Prokop, Ada Huja, Auto-Command, Makis, and Avala.

An assessment of the situation, made in order to create a new plan, indicates only a small degree of realization of the planned solutions. This can be seen in issues affecting both the traffic and the infrastructure: the inadequately developed primary network of traffic, parking systems, and the unresolved issue of Belgrade's nodes; the inadequate use of coastal and aquatic land; an increase in ecological risks; extensive usurpation of agricultural land; intensified illegal construction; uneven spatial development of commercial facilities; intensive multi-annual construction with a reduction in the structure of apartments and the quality of housing; and the loss of identity and importance of the city center.

The adoption of the new Law on Planning and Construction in 2009 resulted in the need for drafting a new General Plan of Belgrade, namely, another harmonization of the existing one. Distinct to the 2003 concept, the Law defined the general urban plan as a strategic development plan with corresponding (reduced) content, level of detail and manner of implementation, which indicated that the GUP of Belgrade had returned to the general planning of the 1970s. However, unlike the time when a plan from the 70s was prepared, this period was characterized by an extremely short production phase and a lack of significant input data and funds for the development of specific studies and expertise. This contradicts the ideals of strategic and long-term planning and the careful consideration of the development of the city. The General urban plan of Belgrade (2016) has identified locations with the potential to generate the development of a wider environment, recognized as future major urban projects directed at urban transformation and reconstruction, including the activation of routes along main roads, accessible and social housing, strategic transport projects, and affirmation of the urban identity. Definition of the directions of the city's development in broad terms, the strategic component of the Plan, was kept to a minimum.

The next document of strategic importance is the Regional Plan for the Administrative Territory of the City of Belgrade, adopted in 2004. The Plan states as its main goal the protection, spatial planning, and development of the city in correlation with its metropolitan area on the principles of sustainable development. It is expected that implementation of the plan would increase the attractiveness of the city and provide the conditions for reaching the standards of other European capitals.

8.2 Visions Determined Through Strategies

The City of Belgrade Development Strategy, the first document of its kind, was prepared in 2008 and adopted in 2011. It was produced by the Palgo center, who gathered experts from different fields to deal with individual sectors. The basic goal of the Strategy (Palgo Center 2011) was to find the best ways in which to raise the City of Belgrade to a higher level, the level of big European cities, by means of a modern economy and advanced technology, so as to help it become a capital in which the citizens will live prosperously, safely, and in a healthy environment. The task of its objectives and strategic priorities is to strengthen the city's position in the constellation of European capitals, i.e., to enhance the economic and cultural structures, to regulate urban and rural areas according to contemporary ecological and aesthetic standards, to integrate natural and cultural heritage that is in its territory, and to highlight the important markers of identity. Considering that the Strategy was planned to cover a period of 4 years, in 2016, the procedure for a new one was initiated.

Although Belgrade needs an integral and long-term strategy, it has also started to define development goals for each sector over a 4-year period. The new City of Belgrade Development Strategy 2021 was adopted in 2017. The authors of the strategy, Regional Development Agency and European Integration (RAREI) and the Palgo center, state that it "precisely defines the roadmap for its long-term sustainable development and allows the capital of Serbia to compare and compete with other metropolises in the region of Southeast and Central Europe, while clear strategic goals determine Belgrade's position in relation to global and European development priorities" (RAREI and Palgo center 2017). However, the vision of an "urban, sustainable, responsible, dynamic and developed European region" does not point to something unique and progressive but continues to move within general terms, as well as solving problems. The strategy defines eight objectives: a competitive economy based on knowledge and innovation, smart urban development of the city on two European rivers, sustainable mobility for interactive development, better and more economical services for all citizens, energy resources as a development opportunity, a sustainable, resilient and renewable city, improved coherence of the economy, and a smart government and administration. After a whole year of employing the strategy, citizens have so far only participated in public insight. As justification for this, it is said that the strategy should serve as a guide to people from the government and administration.

In order to achieve the stated goals, or to follow their implementation, the strategy defines indicators for every objective. In the domain of urban development, the indicators encompass: establishment of the standards for public space regulation; formation of the GIS of the City of Belgrade by 2018; adoption of 60 new regulatory plans by 2021; a minimum 5 objects of culture improved and reconstructed by 2021; and a better position of Belgrade on ranking lists of cities according to the quality of life (RAREI and Palgo center 2017). Indicators of the goal related to sustainable mobility for interactive city development include an increase in the share of pedestrians and cyclists as participants in traffic by 2–5% in the total visual distribution; 100 km of new biking trails built by 2021; established preconditions for the construction of the first line of the metro by 2021; an institutionalized and defined concept of city logistics development by 2021; and improved security conditions for all traffic participants.

Since the adoption of this strategy, it has been almost a year and a half. If we isolate the field of urban development, nothing significant has happened in this domain. The adopted Catalog of urban equipment from 2016 has not yet been supplemented with the key rulebook for public space design. The GIS of the City of Belgrade has been regulated to cover the field of planning documents, but other significant spatial information is not included, which would show a complete picture for individual locations. Not even the development of a smart urban development strategy has begun. With regard to the adoption of plans, it may be considered that this is running at the expected pace, but we will see to what extent the goals are achieved by 2021. If we exclude the reconstruction and reopening of the Museum of Contemporary Art and the National Museum, all other initiatives relating to the restoration and construction of cultural objects (Museum of the City of Belgrade, City Gallery, Philharmonic, etc.) are still on paper. Having a better position in the city ranking in relation to the quality of life is difficult to assess, because other problems, both at the state and at the city levels, such as the rule of law, citizen participation, and corruption affect the tracking and outcome of this indicator.

The General urban plan of Belgrade from 2016 and the City of Belgrade Development strategy from 2017 are documents that determine how the capital will be developed in the near future (because both documents were scheduled for the period up to 2021). The city government representatives expect these documents to define the goals related to relocating industrial zones from the city center, lowering Belgrade to its rivers, and accepting environmental hierarchies in traffic (Folić 2018). Success of the first objective will allow the construction of densely populated residential and business complexes in former industrial areas and the development of the city within its natural boundaries. Achievement of the second goal will be based on removing obstacles to the rivers and furthering direct access to the coast through the construction and introduction of new facilities. The introduction of a green hierarchy in transport will give pedestrians, cyclists, and public urban transport priority, while the city will be built on the principles of the human scale.

8.3 Visions Defined by Public Policy Documents and Initiatives

The current period is characterized by Belgrade undergoing a major urban transformation that includes a number of projects whose implementation will create a "distinctive, attractive and competitive city" (Vukmirović 2016). These projects differ in relation to their character, size, and level of realization and are divided into four main groups: urban plans; traffic, transport and communal infrastructure; investment projects; and improvements in the urban appearance of the city. They are interrelated to form a wheel (Folić 2018). This wheel represents the overall efforts of the city administration to devote equal attention to both grandiose and small-scale projects, with a desire to improve the quality of life of all its citizens and visitors. In this way, Belgrade faces the challenges imposed by the free market and investor urbanism (Čukić et al. 2015), which it was not possible to define more rigorously in planning regulations (Fig. 8.2).

The main element of market-led urban development is seen in the concept of megaprojects based on a top-down planning approach (Zeković and Vujošević 2018), in which decision-making is exclusive and elite-driven. This produces, among other things, the exclusion of local communities from decision-making and

Fig. 8.2 City of Belgrade development wheel (Vukmirović 2016)

insufficient information sharing about future activities, as well as confusion about who can be trusted—politicians, the mayor, bureaucrats, foreign investors, managers, local NGOs, or experts (Zeković and Vujošević 2018). Characterizing it as a "proto-democratic" system, certain authors (Perić 2018; Zeković and Vujošević 2018) state that it has no official limits for citizens—"they are free to express their own opinion, but this is where the story ends" (Perić 2018). On the other hand, Perić (2018) also considers participation as "pseudo-tokenism" in which governing structures manifest power together with "foreign investors capable of achieving their own private interest due to ad hoc spatial planning procedures." In a system that works in such a way "the experts are disabled to fight for their own, while citizen participation is intentionally kept as the lowest priority" (Perić 2018).

The next document was made in the period between the two development strategies, when the Office of the Director of Urban Planning of the City of Belgrade created Project IME: Identity_mobility_environment (Folić and Vukmirović 2016) by analyzing the urban characteristics, open public spaces and quality of life of citizens and considering the current world trends in the field of

Fig. 8.3 Project IME—Identity_Mobility_Environment (Folić and Vukmirović 2016). List of subprojects defined by the document in the period between two strategies

urban planning and design, transport, protection and enhancement of the natural and environmental surroundings and social development of the City of Belgrade. It is a document of strategic character that defines specific thematic frameworks for future interventions, with the aim of achieving three primary objectives: to strengthen the identity of the city of Belgrade, to promote and apply the concept of sustainable mobility, and to enhance environmental awareness. The project encourages the city administration to devote equal attention to both grandiose and small-scale projects in order to improve the quality of life of its citizens and visitors. In line with this, the IME project comprises 20 subprojects (Fig. 8.3), the implementation of which will enable establishing standards and rulebooks for open space management; the use of smart technologies and new methodologies in different domains; establishing a new, environmentally acceptable hierarchy of traffic participants; restoration of facades and improvements in the city lighting; improving the quality of open public spaces, markets, and urban pockets; developing areas of historic and cultural significance; removal of illegal buildings; improving the existing and building new infrastructure; and defining and adopting new strategic and visionary documents that will guide the development of the city over the next 40 years.

In accordance with this document, several initiatives and projects have been completed or are in the process of implementation, such as the Catalog of urban equipment, extension of the pedestrian zones and improvement of local public spaces, introduction of new bicycle paths along the renewed streets and boulevards, reconstruction of the facades, removal of illegal buildings from public spaces, etc. However, this project has failed to start the process of developing a long-term vision for the city's development, apart from a number of initiatives. One of them is the development of Belgrade as a smart city, supported by the introduction of modern technologies and smart government and administration. According to

Cohen's (2015) distinctions of a smart city, the vision of Belgrade as a smart city belongs to the first and second generations of smart cities. Although there are three generations of smart cities—technologically driven, technologically oriented, and citizen-oriented—they do not have to develop linearly from phase to phase, but the cities themselves choose the scenario that suits them best. But the vision of a smart city of the third generation is not easy to accomplish, because Belgrade lacks the development of a very important component, perhaps even a key one, which relates to citizen participation in the planning process (Perić 2018). However, in order to become a truly "smart city," it is necessary to not be just technologically based but to carefully develop and nurture all three components—technologies, communities, and people—i.e., artificial, collective, and human intelligence (Vukmirović 2015). Accordingly, modern technology should be treated as a tool for the purpose of building a better long-term quality of life (Fig. 8.4).

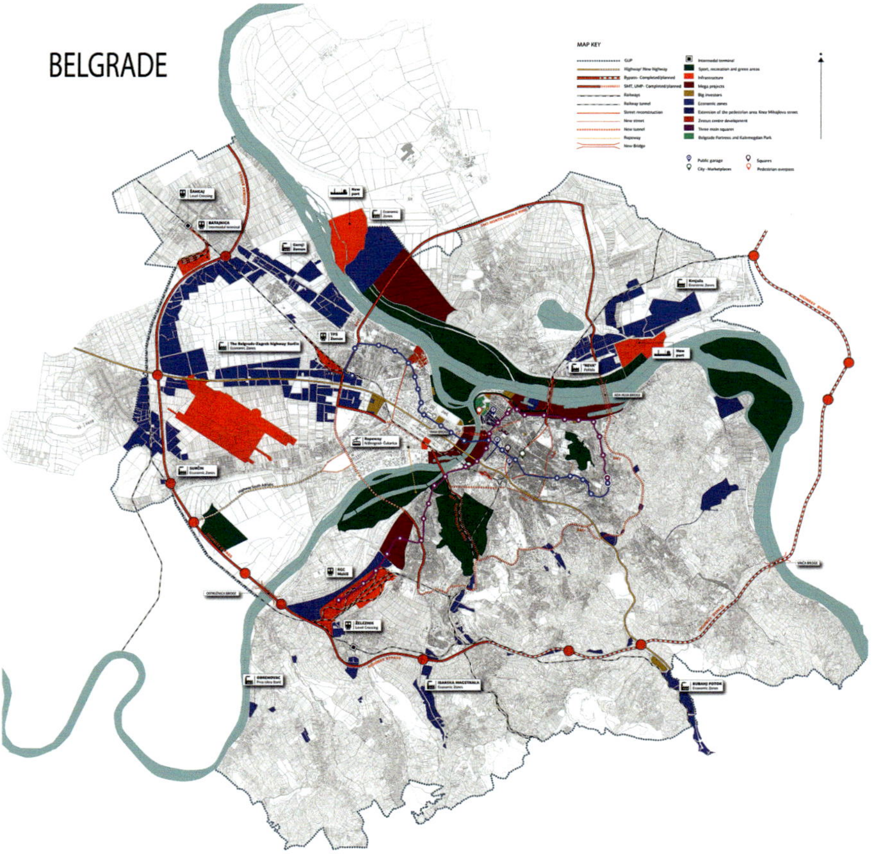

Fig. 8.4 Map of actual and future projects delivered by the Office of the Director of Urban Planning of the City of Belgrade. Author: Ana Delipara 2018

The other development initiative is somewhat broader and re-examines the possibilities of advancing Belgrade as a green capital (Radojičić 2019). In line with this idea, and in recognition of the European cities that have been granted the status of Green Capital of Europe in recent years, the current city government has expressed a desire to participate in this process and submit an application for candidacy. The reason for this is seen in the clear methodology of the candidacy process. It includes an overview of the past, present, and future status of the city in relation to 12 criteria (European Commission 2019): mitigation of climate change, adaptation to climate change, sustainable urban mobility, sustainable land use, nature and biodiversity, air quality, noise, waste, water, green growth and eco-innovation, energy performance, and governance. Each of the above elements must be equally considered and documented by the relevant data, while the methodology itself will serve to define the framework of public policies in the city in the coming period. Aware that they have set an extremely ambitious goal for themselves, the representatives of the city administration are working hard to improve the situation according to each of the defined criteria, in order to improve the quality of life of their citizens. In this way, Belgrade will determine the further course of its own development, and everything that is done in the city will be subordinated to this overall goal of being a green capital with "a consistent record of achieving high environmental standards, to be committed to ongoing and ambitious goals for further environmental improvement and sustainable development and act as a role model to inspire other cities and promote best practices to all other European cities" (European Commission 2019).

Belgrade's other developmental preferences are oriented toward tourism and increasing its attractiveness as a tourist destination (Institute of Economic Sciences 2008). The need for the identification and definition of spatial units (zones and locations) that contribute to the development of city tourism and constitute an important element of the concept of urban development and the enhancement of the identity of Belgrade has resulted in the Plan of Tourist Zones and Locations in the City of Belgrade (Joković Potkotnjak 2017; Urbanistički zavod Beograda 2017). The plan has identified potential areas for the development of tourism; established criteria for defining, and then mapping and spatially determining tourist zones and locations; defined the conditions for arranging and improving existing locations; established new zones and locations; and recognized a basis for passing city decisions and regulations in the field of tourism and hospitality. However, from today's perspective, the development of tourism mainly serves as justification for implementing a number of projects that can only be characterized as city attractions, and which have met with disapproval from the wider, and especially the professional, public. These are projects like Belgrade's gondola that will connect Belgrade Fortress with the Confluence park, the panoramic wheel in New Belgrade, the inventions of Nikola Tesla as an installation in Confluence park, a 120-m spar for the Serbian flag on Savski Square and several 30-m spars for other locations, a fountain on Slavija Square, an integral information system for pedestrians with vague maps and text in Chinese, Russian and English, etc. The attention of the local, but also world public (The Guardian 2018) has been attracted by the New

Year lighting, which is put up in September and removed in March. The New Year lighting installations have been set up without any previously defined concept or project and include catalog-selected elements with the most casual look and inexpensive finishing and effect. Such interventions, although small in size, can contribute to the reputation of the city and must therefore be carefully thought out and be part of a wider vision of the city.

8.4 Conclusion

Based on the above, it can be concluded that Belgrade needs more than one comprehensive, integral strategic document that will determine its long-term development for a period of 30 or 40 years. The General urban plan of Belgrade adopted in 1972 is witness to the fact that this is possible, and it still remains a key developmental document, although it is almost half a century old. Whether Belgrade will continue to develop as a smart city, a green city, or one with some other character, depends on the readiness of the administration, professionals in this area, and citizens to make such a decision. In this way, further deterioration of its identity and the negligent use of natural and territorial resources will be prevented, and the city will be able to develop in the interests of its citizens. In addition, the city will be able to cope with the new challenges that present themselves without much deviation from its defined course. In order to do this successfully, it is necessary to seriously review the current urban practices and to harmonize them with contemporary trends and needs. As Vujosević and Nedović (2006) state, this should be done by the establishment of "a new planning and policy model that (a) is compatible with and supportive of the development of a civic society, and (b) harmonizes with the European Union rules."

References

Arandelovic B, Vukmirovic M, Samardžic N (2017) Belgrade: imaging the future and creating a European metropolis. Cities 63(2017):1–19

Cohen B (2015). The Three Generations of Smart City. https://www.fastcompany.com/3047795/the-3-generations-of-smart-cities. Accessed April 5 2016

Čukić I, Sekulić D, Slavković Lj, Vilenica A (2015) Report from Belgrade Waterfront in Eurozine. https://www.eurozine.com/report-from-belgrade-waterfront/. Accessed online 5 July 2019

European Commission (2019) European green capital. European Green Capital Secretariat. (Online) http://ec.europa.eu/environment/europeangreencapital/about-the-award/

Folić M (2018) Kontinuitet razvoja Beograda. Beograd. Grad Beograd. (Online) http://www.beograd.rs/g/beoinfo/1754702-kontinuitet-razvoja-beograda_2/ (City of Belgrade development continuity)

Folić M, Vukmirović M (2016) Projekat IME - Identitet_Mobilnost_Ekologija. Beograd. Gradska uprava - Sekretarijat za informisanje, Grad Beograd (Project IME-Idenity_Mobility_Environment)

FoNet (2019) Radojičić u Moskvi predstavio projekte i ukazao na izazove razvoja Beograda in Danas. Published online on 6th July 2019. Accessed on 10th July 2019 (Radojičić in Moscow presented projects and pointed out the challenges of Belgrade's development)
Generalni urbanistički plan 2021 (2003) Beograd. Urbanistički zavod (General urban plan of Belgrade 2021)
Generalni urbanistički plan 2021 (2016) Beograd. Urbanistički zavod (General urban plan of Belgrade 2021)
Generalni urbanistički plan Beograda (1950) Beograd. Urbanistički zavod Beograda (General urban plan of Belgrade 1950)
Glogorijević Ž, Graovac A (eds) (2018) 70 Godina Urbanističkog zavoda Beograda Knjiga II: Planovi. Beograd. Urbanistički zavod Beograda (70 years of Belgrade Urban Planning Institute II: Plans)
Institute of Economic Sciences (2008) Strategija razvoja turizma grada Beograda. Beograd. Gradska uprava - Sekretarijat za privredu, Grad Beograd ((2008) City of Belgrade tourist development strategy)
Joković Potkotnjak M (2017) Plan turističkih zona i lokacija grada Beograda. Magazin INFO XV (41):28–32 (City of Belgrade tourist zones and locations plan)
Momčilović V (1974) Beograd u 21. veku. Beograd. Radio televizija Beograd. (Online) www.youtube.comwatchv=LcKn3DoJaGU&fbclid=IwAR1KfzG7au6fbT_Rq1qmr_sCDO-EmBI-B7W6UhVWCbLQKDr8tEoDgscgEesk (Belgrade in the 21st century)
Palgo centar (2011) Strategija razvoja grada Beograda. Beograd. Grad Beograd (City of Belgrade development strategy)
Perić A (2018) Citizen participation in transitional society: the evolution of participatory planning in Serbia. https://doi.org/10.13140/rg.2.2.28467.27686
RAREI Palgo centar (2017) Strategija razvoja grada Beograda: Strateški ciljevi, prioriteti i mere održivog razvoja do 2021. Beograd. Grad Beograd (City of Belgrade development strategy)
The Guardian (2018) Row over Belgrade's Christmas decorations leaves city more furious than festive. https://www.theguardian.com/cities/2018/dec/18/row-over-belgrade-christmas-decorations-leaves-city-more-furious-than-festive
Urbanistički zavod Beograda (2017) Plan turistickih zona i lokacija u Gradu Beogradu (Plan of Tourist Zones and Locations in the City of Belgrade)
Vukmirović M (2015) Belgrade: the quest for the desired city image. In: Planning capital cities: Belgrade, Bucharest, Sofia. Verlag der Technischen Universitat Graz, Graz, pp 188–211
Vukmirović M (2016) City of Belgrade development projects: review. Belgrade. Office of the Director of urban planning of the City of Belgrade (Digital presentation)
Vujošević M, Nedović-Budić Z (2006) Planning and societal context—the case of Belgrade, Serbia. In: Tsenkova S, Nedović-Budić Z (eds) The urban mosaic of post-socialist Europe. Contributions to economics. Physica-Verlag HD, Heidelberg, pp 275–294
Zeković S, Vujošević M (2018) Construction land and urban development policy in Serbia: impact of key contextual factors. In: Bolay JC, Maričić T, Zeković S (eds) A support to urban development process. EPFL, IAUS, Lausanne/Belgrade, pp 29–58

GPSR Compliance

The European Union's (EU) General Product Safety Regulation (GPSR) is a set of rules that requires consumer products to be safe and our obligations to ensure this.

If you have any concerns about our products, you can contact us on ProductSafety@springernature.com

In case Publisher is established outside the EU, the EU authorized representative is:

Springer Nature Customer Service Center GmbH
Europaplatz 3
69115 Heidelberg, Germany

Batch number: 10615512

Printed by Printforce, the Netherlands